Primary Brain Tumors

William S. Fields
Editor

Primary Brain Tumors

A Review of Histologic Classification

Springer-Verlag
New York Berlin Heidelberg
London Paris Tokyo Hong Kong

William S. Fields
University of Texas
M.D. Anderson Cancer Center
Houston, TX 77030, USA

Library of Congress Cataloging-in-Publication Data

Primary brain tumors : a review of histologic classification / William S. Fields, editor.
p. cm.
Based on a meeting held in March 1988 at the University of Texas M.D. Anderson and Cancer Center in Houston.
Bibliography: p.
Includes index.
(U.S. : alk. paper)
1. Brain—Tumors—Classification—Congresses. 2. Brain—Tumors—-Histopathology—Congresses. I. Fields, William S. (William Straus), 1913–
[DNLM: 1. Brain Neoplasms—classification—congresses. 2. Brain Neoplasms—pathology—congresses. WL 15 P952 1988]
RC280.B7P73 1989
616.99'281—dc20
DNLM/DLC
for Library of Congress 89-11363

Printed on acid-free paper.

Camera-ready copy prepared by the author.

9 8 7 6 5 4 3 2 1

ISBN-13: 978-1-4612-8199-3 e-ISBN-13: 978-1-4612-3676-4
DOI: 10.1007/978-1-4612-3676-4

This volume is dedicated to the late KLAUS-JOACHIM ZÜLCH, a man of many talents and boundless energy; clinician, investigator, scholar, teacher, historian, author and administrator. Those of us who knew him well envy his good fortune to have been actively engaged in the pursuit of his chosen work until the very end.

FOREWORD

This volume represents the formal presentations and discussions which took place during a three-day meeting in March 1988 at The University of Texas M. D. Anderson Cancer Center in Houston.

It is dedicated to my friend of more than thirty years, Prof. Dr. Klaus Joachim Zülch, who died in Berlin on December 2, 1988 while this volume was still in preparation. Klaus Zülch had devoted a significant portion of his professional life to a better understanding of central nervous tumors. Over the past two decades he served as the Director of the Collaborating Center for CNS Tumors, under the auspices of the World Health Organization (WHO), and it was largely through his efforts that the work of the Center in developing criteria for a histologic classification of these neoplasms was kept alive. Without his stimulus this Houston meeting would probably not have taken place.

In early 1987 he approached me with the idea of convening, at an early date, a meeting in Houston in collaboration with the Department of Neuro-Oncology of the Cancer Center, of which I was then Chairman. The purpose of this proposed meeting was to discuss recent research developments that might have a profound influence on the classification of brain tumors and ultimately necessitate revision of the "Blue Book" of the WHO on *Histological Typing of Tumours of the Central Nervous System*. The invitees were to be those members of the Collaborating Center who were still active and other neuropathologists from around the world who had expertise in these matters. Support for this project was obtained from The University of Texas endowment funds and from private donations to the Various Donors Account of the Department of Neuro-Oncology.

I thought it would be particularly appropriate to have such a meeting in Houston because of, first, my close friendship with Klaus Zülch over many years and, second, his personal family association with the State of Texas.

He came from a Hessian family and had for a long time been keenly aware that in the 1830's the then Duke of Hesse bought a large tract of land from the Indians in the vicinity of what are now the towns of Fredricksburg and New Braunfels northeast of San Antonio. Subsequently, after the Revolution of 1848 in Europe, many Hessians settled in this region. One of them, Julius Zülch, the brother of Klaus Zülch's grandfather, turned out to be an independent sort of fellow, who after arriving by ship in Galveston, which was then a very important port of entry to Texas, decided to stake out a claim to land further east, in Madison County north of Harrisburg (now Houston). There he established a small community which still exists as the Town of North Zulch. The town still flourishes with two grocery stores, three gasoline service stations, a post office, four churches and a fire station which doubles as a senior citizens' center and domino hall. The railroad station survived until 1986 when it was demolished by a switching engine which jumped from the tracks. Of his many descendents, some of whom still live in the surrounding area, only the widow of Julius' grandson, Jack Zulch, still carries the family name.

I am very much in the debt of my collaborators in this effort, Drs. Dawna L. Duncan Armstrong, Janet M. Bruner, Kenneth M. Earle, Joel B. Kirkpatrick and W. K. Alfred Yung, who helped with the planning and conduct of the program as well as with some of the editorial chores. Special thanks go to my secretary, Mrs. Dorothy Darilek Butler, for her tireless efforts in the preparation of the manuscript in order to have it "camera ready" for printing. I would also like to express my appreciation to the staff of the Conference Services Department of The University of Texas M. D. Anderson Cancer Center for their assistance in the meeting arrangements.

W.S.F.

CONTENTS

PARTICIPANTS

ARMSTRONG, Dr. Dawna L. Duncan, Department of Pathology, Baylor College of Medicine, One Baylor Plaza, Houston, Texas 77030, U.S.A.

BARNARD, Dr. Robin, Department of Pathology, Maida Vale Hospital, London W9 1TL, England.

BECKER, Dr. Laurence, Division of Neuropathology, The Hospital for Sick Children, 555 University Avenue, Toronto, Ontario M5G 1X8, Canada.

BIGNER, Dr. Darell, Preuss Laboratory for Brain Tumor Research, Department of Pathology, Duke University Medical Center, Durham, North Carolina 27710, U.S.A.

BRUCHER, Prof. Dr. Jean-Marie, Neuropathology Department, University of Louvain, Avenue Mounier 52, 1200 Brussels, Belgium.

BRUNER, Dr. Janet M., Department of Pathology, Box 85, The University of Texas M. D. Anderson Cancer Center, Houston, Texas 77030, U.S.A.

BURGER, Dr. Peter C., Department of Pathology, Duke University Medical Center, Durham, North Carolina 27710, U.S.A.

DAUMAS-DUPORT, Dr. Catherine, Department of Pathological Anatomy, Hôpital Sainte Anne, 1 Rue Cabanis, 75014 Paris, France.

DAVIS, Dr. Richard L., Department of Pathology, University of California School of Medicine, San Francisco, California 94143-0506, U.S.A.

EARLE, Dr. Kenneth M., Our Lady of Lourdes Hospital, Lafayette, Louisiana 70501, U.S.A.

FIELDS, Dr. William S., Department of Neuro-Oncology, Box 211, The University of Texas M. D. Anderson Cancer Center, Houston, Texas 77030, U.S.A.

GILLES, Dr. Floyd, Department of Pathology, Children's Hospital of Los Angeles, 4650 Sunset Boulevard, Los Angeles, California 90054, U.S.A.

HASSOUN, Prof. Jacques, Department of General Pathology and Neuropathology, Faculty of Medicine, 27 Boulevard Jean Moulin, 13385 Marseille Cedex 5, France.

ISHIDA, Prof. Yoichi, Department of Pathology, Gunma University School of Medicine, 3-39-22 Showa-machi, Maebashi-shi, Gunma-Ken, Japan.

JÄNISCH, Prof. Dr. Werner, Pathologisches Institut der Humboldt-Universität, Schumannstrasse 20/21, Postfach 140, DDR-1400 Berlin, German Democratic Republic.

JELLINGER, Prof. Dr. Kurt, Ludwig Boltzmann Institute of Clinical Neurobiology, Lainz Hospital, Wolkersbergenstrasse 1, 1130 Vienna, Austria.

KEPES, Dr. John J., Department of Pathology, University of Kansas Medical Center, 39th and Rainbow Blvd., Kansas City, Kansas 66103, U.S.A.

KIRKPATRICK, Dr. Joel B., Department of Pathology, The Methodist Hospital, Mail Station 205, 6565 Fannin, Houston, Texas 77030, U.S.A.

KLEIHUES, Prof. Dr. Paul, Institute of Pathology - Neuropathology, University of Zürich, Schmelzbergstrasse 12, CH-8091 Zürich, Switzerland.

RORKE, Dr. Lucy B., The Children's Hospital of Philadelphia, 1, Children's Center, Philadelphia, Pennsylvania 19104, U.S.A.

RUBINSTEIN, Dr. Lucien, Department of Pathology, Division of Neuropathology, University of Virginia School of Medicine, Charlottesville, Virginia 22908, U.S.A.

SCHEITHAUER, Dr. Bernd W., Department of Laboratory Medicine and Pathology, Mayo Clinic, Rochester, Minnesota 55905, U.S.A.

SOBIN, Dr. Leslie E., Head, WHO Collaborating Center for International Histological Classification of Tumors, Armed Forces Institute of Pathology, Washington, DC 20306, U.S.A.

SZYMAS, Dr. Janusz, Department of Clinical Pathomorphology, Academy of Medicine, 49 Przybyszewskiego Street, 60355 Poznan, Poland.

WECHSLER, Prof. Dr. Wolfgang, Department of Neuropathology, University of Düsseldorf, Moorenstrasse 5, 4000 Düsseldorf 1, Federal Republic of Germany.

YUNG, Dr. W. K. Alfred, Department of Neuro-Oncology, Box 118, The University of Texas M. D. Anderson Cancer Center, Houston, Texas 77030, U.S.A.

ZÜLCH, Prof. Dr. Klaus-Joachim, Max-Planck Institute for Brain Research, Ostmerheimerstrasse 200, 5000 Cologne-Merheim, Federal Republic of Germany.

INTRODUCTORY REMARKS

I. LESLIE E. SOBIN, *Head, World Health Organization Collaborating Center for International Histological Classification of Tumours, Armed Forces Institute of Pathology, Washington, DC*

The WHO *Histological Classification of Tumours of the Central Nervous System* is part of the International Histological Classification of Tumours, a project supported by WHO for the past 30 years. The aim of the project is to provide a standardized system to facilitate communication between cancer researchers and practitioners worldwide.

The classification has been oriented towards defining and naming entities based on images obtained by light microscopy. It recognizes, however, that concepts may change; that they deal more with histogenetic than descriptive approaches; and that they may even be stifled by standards. By separating descriptive terms from conceptual terms, one can avoid controversies that may stand in the way of better comparability of data.

The use of conventional light microscopy has been the mainstay of the WHO classification in order to achieve general application on a worldwide basis. However, immunohistochemistry and electron microscopy are recognized as being able to play an adjuctive role especially when dealing with poorly differentiated tumors.

II. KLAUS-JOACHIM ZÜLCH, *Director, World Health Organization Collaborating Center for Histological Classification of Central Nervous System Tumours, Max-Planck Institute for Brain Research, Cologne-Merheim, FRG*

The 1979 issue of *The Histological Typing of Tumours of the Central Nervous System* was predicated on the state of knowledge at that time. It was foreseen that in future years, after more experience had been accumulated, modifications would be necessary. The classification designed and adopted by the members of the Collaborating Center was a compromise, and of necessity represented some views which differed from those of other participants. It was hoped, however, that for at least a decade, all pathologists would use the classification and if they so desired, publish critiques and suggestions for improvement, simultaneously communicating any critical or dissenting opinions to the WHO or the Center.

A substantial number of concurring papers appeared after this classification had been used for a relatively short time. On the other hand, in the course of time, critical comments were published in medical journals with presentation of varying patterns and concepts of classification, but unfortunately, without communication to the WHO or to the Center.

As a result of these preliminary activities, those leaders of the Collaborating Center who remained active, were invited, at a meeting in Vienna in 1986, to continue their activities by promoting the acceptance of the 1979 issue of the *Blue Book* by various neurospecialties. Arrangements for convening meetings of interested scientists in satellite workshops at national or international congresses were encouraged.

For further refinement, the prerequisites to be applied for classification were defined as follows:

1. Classification should be based on the microscopic characteristics of tumors, e.g. the morphologically identifiable cell types and histologic patterns seen with conventional light microscopy.
2. Time-honored and widely-used terms should be retained unless considered misleading or undesirable (where synonyms had been added in brackets for better understanding, the preferred terms should remain marked); the suggestion that for purposes of identification, light microscopy should be used, applies only to daily routine diagnosis, whereas more sophisticated methods now available should be adopted only to explain the remaining scientific tumor problems but within the existing classification.

The plan to describe the morphology in the simplest terms possible and by methods applicable to daily light microscopic routine was to continue. This goal would remain unchanged although consideration was to be given to immunohistologic and other marker methods which have meanwhile been introduced into scientific investigations. These investigations might eventually provide more information concerning the origin and position of some tumor groups within the classification of nervous system tumors.

At this meeting in Houston an effort was made to determine whether changes in the basic classification outline would be necessary, keeping in mind that the value of the *Blue Book* had already been proven in daily routine work. Classification must remain clear, informative, as simple as possible and only as detailed as necessary. The present classification should be changed only when the text and the statements no longer correspond with generally and overwhelmingly accepted opinions and when change clearly would bring advantage to the daily work of the pathologist.

PART ONE

EMBRYONAL TUMORS OF THE NEUROECTODERM

Moderators:

DAWNA L. DUNCAN ARMSTRONG

and

KENNETH M. EARLE

CHAPTER I

Primitive Neuroectodermal Tumor – A Concept Requiring an Apologia?

LUCY BALIAN RORKE, *The Children's Hospital of Philadelphia and University of Pennsylvania School of Medicine, Philadelphia, PA*

Introduction and Historical Review

Concepts relating to cell of origin and nosology of the group of central nervous system (CNS) tumors composed largely of primitive neuroepithelial cells (PNECs) date back to 1925 when the prototype tumor in this group, the medulloblastoma, was defined as a clinical and pathologic entity by BAILEY and CUSHING (1925). Tumors of this type had been described by others around the turn of the century but had carried a variety of names, most including the term *sarcoma* (ZÜLCH 1965).

Bailey identified 29 tumors belonging to this group from among approximately 400 surgical specimens removed by Cushing. Frustrated by lack of interest of the pathologists at the Peter Bent Brigham Hospital, Cushing had assigned Bailey, then one of his young trainees, to establish a neuropathology laboratory and classify the specimens (BAILEY OT 1985-86).

Utilizing histogenetic *cell of origin* concepts advanced by two German pathologists, E MÜLLER (1903) and RIBBERT (1910), Bailey and Cushing initially asserted that the tumors were composed of spongioblasts, neuroectodermal cells regarded as progenitors of glia, and named the tumor *spongioblastoma cerebelli*. However, they discarded this name almost immediately because GLOBUS and STRAUSS (1925) had simultaneously adopted the term *spongioblastoma* for the tumor we now call *glioblastoma multiforme* (BAILEY and CUSHING 1925). How different the medical literature would have been but for this chance event! Because of this, Bailey and Cushing were obliged to rename their tumor. They created the unique name *medulloblastoma*, as they revised their opinion of its origin from spongioblasts and suggested that the tumor resulted from neoplasia of a medulloblast, a hypothetical, multipotential neuroepithelial cell that could differentiate into an astrocyte, oligodendrocyte or a neuron (BAILEY and CUSHING 1926).

They chose to ignore neuroembryologic data advanced by RAMON Y CAJAL (1890a) and HIS (1904), whose studies suggested that the primitive neural tube contained a mixed population of two types of cells, namely neuroblasts and spongioblasts, which are already destined to develop into neurons and glia, respectively.

Instead, they proposed that the neural tube contained five types of stem cells, namely choroidal epithelium, pineal parenchyma, primitive spongioblasts, apolar neuroblasts and medulloblasts (BAILEY and CUSHING 1926). They apparently misinterpreted SCHAPER's concept (1897) that all cells in the ventricular or the germinal zone consisted of a monomorphic population of *indifferent cells* that were multipotential and could differentiate into a neuron or glial cell. Furthermore, whereas Schaper thought that the *indifferent cell* occurred ubiquitously in the developing nervous system, KERSHMAN (1938) later claimed that it was unique to the cerebellum and was, in reality, the external granular cell. He further asserted that tumors arising from them, namely, Bailey and Cushing's medulloblastomas, were therefore unique to the cerebellum.

Although it was generally agreed that tumors of this histologic type were most common in the cerebellum of children, CUSHING (1930) and BAILEY (1948) (grudgingly) noted that they occurred in the cerebrum and spinal cord. BAILEY (1932) even called attention to their histologic similarity to retinoblastomas. However, with the passage of time it became customary to call only the cerebellar examples *medulloblastoma* and to give a different name to histologically similar tumors arising in other sites in the nervous system (HART and EARLE 1973; AZZARELLI et al 1977; OJEDA et al. 1978; PALMER et al. 1981; RUBINSTEIN 1981; HASSOUN et al. 1982).

Arguments in support of the medulloblastoma as a tumor unique to the cerebellum have been forcefully stated by RUBINSTEIN (1972a; 1975) and are briefly summarized as follows:

1) the medulloblastoma cell of origin is a germinative cell derived from the cerebellar external granular layer, possibly from the internal granular cells or from nests of primitive germinal cells in the posterior medullary velum, and
2) the medulloblastoma cell of origin is a primitive neuroepithelial cell capable of two-directional differentiation (RUBINSTEIN 1972a).

Controversial Issues in Classification Based on Histogenesis

These concepts relative to medulloblastoma and other embryonal central neuroepithelial tumors are based upon the histogenetic approach introduced by J MÜLLER (1838) for application to tumors in general, and adopted by BAILEY and CUSHING (1926) for central neuroepithelial tumors. This histogenetic approach was criticized by ROUSSY and CORNIL (1928) shortly after it was proposed, who voiced two major objections: 1) classification based upon histogenesis is hypothetical and unprovable, and 2) neoplasia results from anaplastic trans-formation of normal cells rather than from cells arrested at certain stages of their development. Other critics of this approach included Cox (1933), in years past, and two eminent contemporary pathologists.

ZÜLCH (1965), in the second edition of his book, *Brain Tumors - Their Biology and Pathology* makes the following remark:

> "This histogenetic interpretation has proved to be so defective as a doctrine ... *that there is no justification for its further expansion or the introduction of new terms*. Similarly, to deduce a histogenetic system for normal development by basing it on the findings in tumors seems to be putting the cart before the horse."

GOULD (1986), in a superb essay reevaluating concepts of histogenesis and differentiation as criteria for classification of tumors, offers compelling arguments to base classifications on consistently applied and objectively demonstrable features as defined by a series of appropriate phenotypic differentiation marker expressions, rather than on the basis of questionable assumptions about their origins.

A student faced with conflicting views of respected teachers has several options:

1) to select one view because of the authority of its protagonist; or
2) to accept only those concepts that are based upon proven or provable facts; or
3) to conditionally accept features of one or more views based upon reasonable scientific data, and accumulate sufficient experience to establish a personal view. This third approach may be likened to Szent-Gyorgi's credo: "to see what everyone else has seen, but think what no one else has thought" (Moss 1987).

Although to claim that I "think what no one else has thought" would be impertinent, I boldly challenged established concepts relating to cell of origin and nosology of central embryonal neuroepithelial tumors (RORKE 1983). My object was to draw attention to inconsistencies in nomenclature, and certain tenets that had been shown to be incorrect or were untestable. In so doing, I roiled a hornet's nest.

Critique of Prevailing Concepts

On the basis of my knowledge of the literature, and evaluation of tissues from a large number of children with CNS tumors composed of primitive neuroepithelial cells (PNECs), I distilled five observations that appeared valid:

1) tumors composed of PNECs occur most commonly but not exclusively in the cerebellar vermis of children;
2) such tumors may contain neoplastic cells with features characteristic of astrocytes, oligodendroglia, ependymal cells or neurons. Pigmented cells are seen in exceptional examples;
3) the unique name assigned to the tumor cannot be applied if site of origin is not known;
4) the majority are biologically malignant and have a tendency to disseminate throughout cerebrospinal pathways;
5) the tumors have a variable connective tissue component, and rarely contain smooth or striated muscle.

As a corollary, I drew four additional conclusions:

1) the neural tube does not contain five types of primitive cells as hypothesized by Bailey and Cushing;
2) the cerebellar external granular cell is not bipotential, i.e. it can only develop into a neuron but not a glial cell; and
3) it is impossible to predict differentiating potential or determine ancestry of a primitive neuroepithelial cell (PNEC) in a tumor by currently available morphologic techniques.
4) Analysis of data derived from application of molecular markers to tumors makes it clear that old concepts relative to cell of origin and differentiation potential require critical reexamination.

Examination of each point may assist in clarifying these important issues.

1. The neural tube does not contain five types of primitive cells as hypothesized by Bailey and Cushing.

This hypothesis by Bailey and Cushing is unique, as the majority of neuroembryologists for the past 100 years distinguished one or at the most two separate cell lines to account for the origin of neuronal and glial cells. Among the earliest was His (1904) who suggested two separate cell lines in the germinal matrix (ventricular zone): one group that gave rise to neurons and a second group to spongioblasts. At about the same time, SCHAPER (1897) proposed that the germinal matrix is composed of proliferating indifferent cells that migrate into the intermediate zone where they further divide into either neurons or glial cells. More recently, FUJITA (1963; 1986) suggested that the matrix cells first give rise to neurons and then the same dividing population begins to produce glial cells. In contrast, RAMON Y CAJAL (1890b) noted two cell lines in the matrix zone as reported by HIS (1904), but suspected that one line gave rise very early in development of a special glial cell now recognized as radial glial (RAKIC 1981; LEVITT et al. 1981).

Current evidence suggests that the ventricular zone of the neural tube contains two types of embryonal cells: one that is committed to develop into a neuron and a second type that is destined to give rise to macroglia (RAKIC 1981; LEVITT et al. 1983). This glial progenitor cell, however, is apparently bipotential and gives rise to three types of macroglial cells through two distinct lineages, i.e. the 0-2A progenitor from which oligodendrocytes and type 2 astrocytes develop, and a separate precursor cell whose progeny are type 1 astrocytes (RAFF et al. 1983; MILLER and RAFF 1984; TEMPLE and RAFF 1985; RAFF et al. 1987).

2. The cerebellar external granular cell is not bipotential, i.e. it can only develop into a neuron but not a glial cell.

Speculations regarding the cerebellar external granular cell and its differentiating potential date back to the late 1800's. SCHAPER'S studies (1894) of neuroepithelial cells lining the posterior lip of the fourth ventricular roof led him to believe that these migrating cells, destined to give rise to cerebellar external granular cells, were still undifferentiated but bipotential, and hence he called them *indifferent cells*. Some of his contemporaries (LUGARO 1894; POPOFF 1895) and later investigators (JAKOB 1928) agreed with him, although RAMON Y CAJAL (1890) contended that mature granule cells are the only derivatives of the external granular layer. Schaper subsequently expanded his concept of the indifferent cell and stated that similar indifferent cells played a fundamental role in development of the brain and spinal cord, as already noted above.

However, KERSHMAN'S (1938) studies of the embryonic nervous system led him to conclude that the only identifiable *indifferent cell* in the developing nervous system was the cerebellar external granular cell, and hence this was truly the hypothetical medulloblast postulated by BAILEY and CUSHING (1925).

Investigations by a growing number of contemporary workers have added support to Ramon y Cajal's original claim that the external granular cells are committed as neuroblasts, refuting the suggestion that they also give rise to glia (SWARZ and DEL CERRO 1977; KOPPEL and LEWIS 1978; FÜLOP et al. 1979; GHANDOUR et al. 1981; GHANDOUR et al. 1983). Under the circumstances, there is no justification for clinging to the concept of a nonexistent *indifferent cell*, the mythical medulloblast, or to the notion of a bipotential external granular cell that can whimsically develop into a neuron or a glial cell.

This is not to say, however, that neoplastic transformation of the external granular cell cannot result in formation of an embryonal neoplasm in the cerebellum. It is, in fact, a distinct possibility particularly as cells comprising a mitotically active population (as these are late in gestation and during the first year of life) are at increased risk as targets for carcinogenesis (FRIEND et al. 1988). Moreover, there is no reason why the internal granular cell or the primitive cells of the posterior medullary velum may not also be targets for neoplastic transformation.

If the transformed external or internal granular neuron expresses its normal genotypic destiny, it would most likely form a neoplasm with neuroblastic features, at least in the initial generation. However, it is conceivable that progression of later mutations could lead to loss or abnormal expression of neuronal features and reversion to the embryonic, undifferentiated state, or to acquisition of a totally different phenotype.

3. It is impossible to predict differentiating potential or determine ancestry of an embryonal neuroepithelial cell in a tumor by currently available morphologic techniques.

We define the primitive neuroepithelial cell (PNEC) by various criteria for examination with the light microscope, electron microscope, and with special markers utilizing immunoperoxidase techniques or monoclonal antibodies. With the light microscope these PNECs are said to be small to medium-sized, round to oval cells with basophilic nuclei with or without nucleoli, and scanty cytoplasm. At the ultrastructural level the cells are characterized by nuclei with or without prominent nucleoli surrounded by scanty cytoplasm containing inconspicuous cytoplasmic organelles. Studies utilizing immunocytochemical techniques or monoclonal antibodies disclose no features characteristic of specific, more mature cell types, although some may contain the intermediate filament, vimentin (TROJANOWSKI 1987). However, if glial fibrillary acidic protein (GFAP) or neural filament protein (NFP) are present, the cell can no longer remain in the embryonal category. At the same time, all those tested to date exhibit a pan-neuroectodermal marker (BOURNE et al. 1986; FRIEDMAN et al. 1988).

Sufficient data have not yet accumulated to state whether synaptophysin is yet another pan neuroectodermal marker. This is an integral, acidic membrane glycoprotein of presynaptic vesicles (M 38,000) in various neurons and neuroendocrine cells, and in tumors derived from such cells (SCHWECHHEIMER et al. 1987). These same investigators, using a monoclonal antibody to synaptophysin, observed consistent positive immunoreactivity in six primitive neuroectodermal tumors (PNET)-medulloblastomas and three neuroblastomas. Although NFPs were consistently expressed in the neuroblastomas, intermediate filament distribution in the PNET-medulloblastomas was variable. Four of the six expressed vimentin, but only one of the six contained neural filament (NF) positive cells.

Currently available techniques do not allow us to state whether neoplastic PNECs arose consequent to transformation of a previously normal PNEC or through anaplastic evolution or progression of a more mature cell type. In other words, the only valid statement that can be made about a population of neoplastic PNECs is what they, in fact, are. Speculations regarding their ancestry or differentiating potential are unjustified and irrelevant.

4. Analysis of data derived from application of molecular markers to tumors makes it clear that old concepts relative to cell of origin and differentiation potential require critical reexamination.

There is little disagreement that PNETs may contain malignant astrocytes, oligodendrocytes, ependymal cells, neurons, pigmented cells, smooth or striated muscle, singly or in combination. HERMAN and RUBINSTEIN (1984) reported a remarkable example of organ culture studies of one such tumor arising in the cerebellum in which glial, neuronal and pineocytic features were demonstrated. Application of special markers utilizing immunoperoxidase methods for identification of intermediate filament proteins denoting specific cell types has revealed, however, that these apparently *differentiated* cells are not necessarily normal neurons or glia (TROJANOWSKI 1987). For example, it has been found that neoplastic cells may co-express two intermediate filaments, although normal mature cells with similar aggregates are unknown. Moreover, Trojanowski has observed abnormal arrangements of the neurofilament proteins in the D283MED cell line derived from a PNET-medulloblastoma. These consisted of abnormal perinuclear aggregates not seen in normal neurons, absence of neurofilament L but expression of neurofilaments M and H, and few intermediate filaments in most cells or absence of side-arm projections that are characteristic of normal neurons.

Complicating the problem further is at least one major exception to the rule that neurons can be recognized by their expression of neurofilament proteins, i.e. the sole intermediate filament protein (IFP) of the olfactory sensory neurons is vimentin, not neurofilament (SCHWAB et al. 1986).

The other types of apparently *differentiated* neoplastic cell types that form the spectrum of phenotypes in these tumors may also be deranged. However, the possibility that totally normal neurons, glia, ependyma, etc. may result from neoplastic progression in these tumors cannot be excluded. For example, it is conceivable that the apparently differentiated astrocytes so commonly found in PNETs and considered reactive forms, have, in reality, arisen through a neoplastic process.

While increasing utilization of markers has expanded our knowledge of this group of tumors, it has raised significant interpretive problems. Thus it appears that genetic control and expression of PNECs forming tumors are sufficiently abnormal that attempts to draw parallels between their behavior and those following normal pathways in development of a functional nervous system are invalid.

Findings by FRIEDMAN et al. (1988) of a medulloblastoma cell line nurtured in tissue culture emphasize the complexity of these problems. The original tumor from which their culture preparations were derived was negative for GFAP and NFP. However, one cell line, D341, that grew from the tumor contained two of three subunits of NFP, namely the high and medium molecular weight neurofilament proteins. These cells also reacted positively to a pan-neuroectodermal marker. On the basis of their investigations, the authors concluded that the tumor they studied was a neuroblastic neoplasm. However, evidence of its neuroblastic differentiating potential was not apparent from studies of the original tissue specimen, which, as noted, was negative for NFP. This same group (FRIEDMAN et al. 1985) established another cell line from a medulloblastoma, D283, which on the original biopsy expressed both GFAP and NFP. However, in culture the cells expressed only NFP. BOURNE et al. (1986) report a similar observation in a PNET of the cerebrum that was diagnosed cerebral neuroblastoma but did not express NFP in tissue sections. When placed in standard tissue culture, the cells produced 210K neurofilament antigen recognized by their monoclonal antibody, RT97, after one subculture.

Evidence Supporting Concept of PNET

The majority of practicing pathologists do not have complex tissue culture facilities or an array of monoclonal antibodies available for such elegant analyses of all tumors composed of PNECs but must rely upon those practical techniques that can be used in a routine diagnostic laboratory. On the basis of an ordinary H&E-stained specimen and examination of sections stained for identification of GFAP, NFP, or other appropriate commercially available markers, the only conclusion the pathologist can draw is that the tumor under investigation, in one case, FRIEDMAN's (1988) *neuroblastic* medulloblastoma or BOURNE's (1986) cerebral *neuroblastoma* is composed of PNECs, period. Prediction that they would manifest the neuronal features noted in tissue culture preparations is not possible. A major lesson to be learned from Friedman's and Bourne's studies is that the pathologist can only draw conclusions on the basis of what is seen. He is neither a genealogist nor a prophet.

A noted jurist once said, "Our duty is to believe that for which we have sufficient evidence, and to suspend our judgment when we have not (CRITCHLEY 1986)." This practice in the courts of law should be adopted by all who seek to unravel nature's mysteries. If we apply this principle to the information currently available relative to the controversial issues surrounding the cell of origin and nosology of the central PNETs, we are left with the following evidence:

1. The cerebellum does not contain a unique bipotential germinative cell that can mature into either a neuron or a glial cell. In the absence of such a unique cell arguments supporting the concept of a cerebellar tumor unlike any other type of CNS tumor are untenable.
2. Although most common in the cerebellum, PNETs may arise anywhere within the CNS, either by the process of neoplastic transformation of a previously normal PNEC or through anaplastic evolution or progression of a more mature cell type. It is, therefore, illogical to apply a specific name to a PNET on the basis of location in the CNS. We do not, for example, have one name for meningiomas in the supratentorial space, a second term for those in the posterior fossa or yet a third for this type of tumor when it is located in the spinal canal.
3. A PNET may appear to consist of a monomorphic population of cells when examined by light and electron microscopy and after application of a battery of special markers. However, these cells may carry a genetic code for glial, neuronal or other cell types that may be expressed at a future time. The tumor may, of course, consist of a phenotypically polymorphic cell population at the time of initial examination. It is also possible that evolution or progression of the PNEC through chromosomal rearrangements may produce irreversible changes into a variety of cell types, which

with the passage of time, do not necessarily remain stable and bear little resemblance to the original neoplastic clone. Observations to date confirm the basic neuroepithelial character of the group of PNETs but clearly document their heterogeneity. Detailed phenotypic or even genotypic analysis of a given tumor does not allow prediction of the phenotype or genotype of any other tumor in this general group. In other words, each tumor is unique.

Under the circumstances, prudence dictates that classification of this group of PNETs should be based on consistently applied criteria of objectively demonstrable histologic features. Theories relative to their oncogenesis require validation by rigorous testing in the laboratory. The medical specialty of pathology should be a scientific discipline based upon observable facts and testable hypotheses; it is not a religious sect whose followers accept tenets on faith.

In these matters it may be wise to follow RAMON Y CAJAL's (1937) view:

"I have always felt an invincible repugnance for polemics. Precious time is wasted on them, which could be used better in gathering new facts. Besides, who does not know that truth, even though undefended, will prevail in the end?"

Summary

Controversies dealing with concepts of cell of origin and nosology of the cerebellar medulloblastoma and other histologically similar central nervous system tumors are reviewed. Critical reexamination of disparate opinions by neuroembryologists and neuropathologists utilizing data obtained through sophisticated cell marker and tissue culture studies, lead to the following conclusions:

1) the concept of a uniquely cerebellar medulloblast or other indifferent neuroepithelial cell is untenable;
2) although most common in the cerebellum, tumors composed of primitive neuroepithelial cells may occur anywhere in the central nervous system;
3) currently available tissue markers confirm the neuroectodermal nature of cells forming these tumors;
4) cells comprising the tumors may be primitive or manifest one or more markers associated with *differentiation* although the molecular expression of this *differentiation*

is not necessarily equivalent to normally-developed neural cells;

5) each tumor is unique, such that detailed investigation of one does not permit prediction of the character of another;
6) examination of a given specimen does not allow determination of ancestry or differentiating potential by any currently available technique - the investigator can only delineate the unique features of the material at hand; and
7) prudence dictates utilization of nomenclature for the group of CNS primitive neuroepithelial neoplasms based upon their phenotypic characteristics with a secondary notation of site of origin.

Future advances may allow classification at the genotypic level or force revision of currently acceptable nosology.

CHAPTER II

Justification for a Cytogenetic Scheme of Embryonal Central Neuroepithelial Tumors

LUČIEN J. RUBINSTEIN, *Department of Pathology, Division of Neuropathology, University of Virginia School of Medicine, Charlottesville, VA*

Introduction

The concept of central neuroepithelial tumors of primitive or perhaps, as one should rather call them, embryonal type requires no apologia. The existence of this category of neoplasms has long been recognized, even in the days when the main emphasis was on the view that most tumors of the glioma group presumably arose from already differentiated cells which underwent *dedifferentiation* in the course of growth. Thus WILLIS (1960) singled out the cerebellar medulloblastomas from other glioma types as "truly embryonic tumours arising from still immature neural tissues during foetal life or early childhood." There is widespread agreement that, followed by the retinoblastoma, the cerebellar medulloblastoma is by far the commonest of the central nervous system (CNS) neoplasms in that category. As a result, it is the one whose histogenesis and differentiation potential have been most extensively investigated and discussed.

Despite the apparent cellular homogeneity of this group of tumors in routine histologic preparations, the subject of their classification is complex. This is because we are concerned not only with recognizing their cytologic characteristics, but also with devising an orderly scheme which, to be meaningful, should satisfy four criteria: 1) it should have some logical consistency; 2) it should recognize established and generally accepted clinicohistologic entities (i.e. entities about which there may be disagreement as to how they should be named, but an agreement that they exist and can be consistently identified); 3) it should not cause confusion or misunderstanding to nonspecialists by being so ill-defined in its criteria that it will introduce other tumors in which closely similar cellular features are the result of focal or widespread anaplasia; and 4) it should have clinical and therapeutic relevance, and not result in the application of protocols which, while reasonably successful in some types of embryonal tumor, will be ineffective or disappointing in those examples whose appearances are largely those of anaplastic change. An indiscriminate application of an all-embracing concept that includes all *primitive neuroectodermal tumors* into a single tumor entity is, in my view, a simplistic solution to the problem. It has resulted in a disconcerting number of misdiagnoses on the part of general pathologists and of oncologists untrained in neurohistology. Such a concept also ignores or underplays the fact that the cells of the various entities covered by this term differ in their localization in the CNS and vary in their differentiation potential. The objective of this presentation is to suggest an alternative scheme that proposes, as was indicated many years ago (RUBINSTEIN 1972b), a classification which should preferably be based on a cytogenetic scheme, even though such a scheme may still be imperfect and therefore may in the future have to be modified in the light of more data.

Reasons for the complexity of the problem

Why is the problem complex? The first reason is that neoplastic disease is a process which largely involves cells of renewal, i.e., stem cells which lack the differentiating characteristics of the tissue cells to which they give rise. The stem cells carry only the genetic information for the differentiation of their progeny and differentiation is expressed largely in those mitotic cells which have attained their terminal stage of proliferation (PIERCE et al. 1978). In tissues that normally possess a considerable reserve of stem cells whose progeny is capable of further differentiation (e.g. the skin or the internal epithelia), the distinction between adult-type tumors and embryonal tumors (i.e. tumors which arise during embryonal, fetal or postnatal development from tissues which are still immature, but in which differentiation is already determined and restricted) (WILLIS 1958) is relatively easy. However, in the CNS the distinction becomes blurred because the cells of renewal (therefore the presumed targets for the first hit of neoplastic change) are largely to be found while the neural tissue is still developing. This is a reasonable assumption, based on the evidence that in the adult primate brain the reserve neuroepithelial stem cell population must be either absent or very low and that any existing turnover of glial cells will be slow (KAPLAN 1982; RAKIC 1985). Thus a general concept may be introduced that will perhaps facilitate understanding of the problem and might account for many aspects of CNS tumor incidence and for the relationship of embryonal CNS tumors to tumors of adult cell type. The concept is that a window of neoplastic vulnerability exists which may be either wide or narrow. The width of the window depends on the interaction of several factors, the most important of which is the length of time during which a reserve population of still replicating stem cells may be exposed to the first hit of neoplastic transformation. The width of the window can be correlated with the frequency of the different types of embryonal CNS tumor in man (RUBINSTEIN 1987a).

Important to the concept of a window of neoplastic vulnerability are the presumed state of differentiation of the replicating stem cells at risk and the future differentiation potential of the progeny of those targeted cells. This may be significant in determining the differentiation potential, and also the restrictions in differentiation potential, of the different embryonal tumors that may present in different sites of the central neuraxis. Embryonal tumors display a capacity for differentiation which is similar to that of the parent developing tissue and which usually is less restricted than that of tumors originating from postnatal cells. In addition, neoplastic differentiation often is less limited than normal differentiation. This last feature is not necessarily limited, as might be anticipated, to embryonal tumors, but the capacity for abnormal differentiation (or heteroplasia) has long been recognized to be particularly frequent in tumors of the embryonic category (WILLIS 1958).

A second factor compounds the difficulties of an already difficult problem. This is the frequency with which anaplastic change occurs in any tumor arising in the CNS. It may result in a virtual impossibility, from a simple survey of the morphologic features, to decide whether the poorly differentiated cells are anaplastic or embryonal. Speaking in kinetic terms, anaplasia results from the preferential selection and predominance of clonogenic subpopulations that emerge in a tumor which may originally have been heterogeneous in its karyotypic and phenotypic characters. The dynamic expression of this evolution is an increase of the tumor growth fraction, i.e. cells from the nonproliferating, or quiescent, pool are being recruited into the proliferating cell pool. While this aspect is easily understandable in kinetic terms, it may cause considerable confusion in the cytologic interpretation of embryonal CNS tumors. This is especially the case in those tumors which have been assumed (usually erroneously, in my view) to present in adults. However, the error may also be made in childhood and infantile gliomas. Astrocytic gliomas may undergo anaplastic change at any age, and the resulting emerging subpopulations with proliferating advantage may then form classic glioblastomas or medulloblastomas (RUBINSTEIN 1985b).

Frequency of clinicopathologic entities as related to the window of neoplastic vulnerability

In practice, we are confronted by some eight distinct histopathologic entities (the cerebral medulloepithelioma, the cerebral neuroblastoma, the primitive polar spongioblastoma, the ependymoblastoma, the pineoblastoma, the cerebellar medulloblastoma, the olfactory esthesioneuroblastoma and the retinoblastoma). The frequency of those tumors corresponds quite well to the width (either narrow or wide) of the window of neoplastic vulnerability as we defined it (RUBINSTEIN 1987a). All these entities have been documented from the histopathologic viewpoint, but their biologic behavior still needs to be better known in some of the rarer tumors. These designations are currently being used by most pathologists and neuropathologists; more than 90% of the primitive tumors arising in the CNS fit into one of those diagnostic names and present no serious histologic diagnostic problem. Two of the entities (the cerebellar medulloblastoma and the retinoblastoma) are relatively common and have been extensively studied. The others are rare and more examples are needed to permit more exact clinicopathologic correlations. So far, only very few instances of central neuroepithelial embryonal tumor have not, after careful study, found a place in such a frame of reference. It is in those exceptional cases that further pathologic documentation is required and it is largely in examples of that kind that additional information will probably be provided by modern techniques of investigation.

Pitfalls of a simplistic concept of "primitive neuroectodermal tumors"

The concept of a single diagnostic term *primitive neuroectodermal tumors* (PNET) as an independent entity from which the others derive (with the added qualification of non-differferentiating or of variously differentiating elements) (RORKE 1983; RORKE et al. 1985) has in my view several flaws, and these flaws have had rather negative consequences in the establishment of therapeutic protocols and in the results that were anticipated from their application. The PNET, based on a paper published in 1973 and reporting a series of 23 brain tumors in children (HART and EARLE 1973), is not in my opinion a very convincing single entity. The criteria for its recognition lack precision. Some of the described tumors may have been anaplastic small-cell gliomas, others could represent cerebral neuroblastomas. Usually, however, the documentation is too incomplete to permit their assignment to a definite category. These examples have been qualified as *left-overs* (RORKE 1983) or as representing a *wastebasket* category (RORKE 1986). These are not my terms; they have been used by the proponents of the scheme we are discussing.

That scheme takes no account of the differentiating potential operative in the different cells of the primitive neuroepithelial matrix and at different sites of the CNS, and largely dismisses the part presumably played by local ontogenetic factors in determining the lines of differentiation, and the restrictions of differentiation potential, at those sites. The implication is that all normal primitive neuroepithelial cells are capable of differentiation along any or all cell lines, irrespective of their particular stage of cytogenetic development. The proponents of the scheme are of course aware of this problem and have briefly raised it in their preamble, but have left the question open (RORKE et al. 1985).

A more serious concern is that this scheme makes no distinction between those entities that have been extensively investigated - the cerebellar medulloblastoma and the retinoblastoma - and those which, though often recognized by competent pathologists, are rare and on which more documentation is desirable. Well-established entities and topographies are dismissed; all are placed within the same common denominator.

This oversimplification seems to distort nosologic definitions. In practice, it results in a good deal of misdiagnosis on the part of those less experienced than the proponents of the scheme, in spite of their intention to adopt a terminology that is purely practical (RORKE et al. 1985).

One of the sources of error for the inexperienced lies in the difficulty of distinguishing between anaplastic and embryonal tumors, both of which are largely made up of small, round or oval, undifferentiated cells. A second confusion has originated, as we have mentioned, from the blurring between entities which are already well known and those which are not. In summary, this concept does not seem to advance our understanding or our recognition of a difficult group of childhood tumors.

The negative results are already apparent in the literature and in daily referral practice. Thus, one paper which describes the clinicopathologic correlations of a number of cerebral tumors originating in the supratentorial compartment of children and young adults, and interpreted as PNETs with various lines of differentiation (GAFFNEY et al. 1985) records their lack of correlation with the results obtained in the treatment of cerebellar medulloblastomas. In that series, however, the documentation suggests that many of the tumors, especially those in adult patients with short periods of postoperative survival and in which the microscopic picture was associated with vascular endothelial proliferation and positivity of the tumor cells for glial fibrillary acidic protein, were various forms of anaplastic glioma rather than true embryonal tumors (RUBINSTEIN 1985a). I have now had referred to me many examples, mostly in children, but also occasionally in adults, in which a diagnosis of PNET was entertained following its introduction in the literature, but which, on more careful study, almost invariably turned out to be another entity: a cerebral neuroblastoma, a small-cell infantile glioma, a glioblastoma with a largely monomorphic small-cell population, an anaplastic astrocytoma, a malignant oligodendroglioma, or a cerebral metastasis from a previously diagnosed cerebellar medulloblastoma. A terminology which results in confusion of this extent cannot, I believe, be regarded as a firm foundation on which to establish an orderly scheme of classification.

The interpretative problems become especially difficult in the case of very small fragments obtained from a limited biopsy and which as a consequence may not be representative of the rest of the tumor or cannot be studied by techniques more sophisticated than routine stains. It could perhaps be argued that any histologic diagnosis is better than none because of the necessity of making a therapeutic decision even if the latter is based on limited or inadequate tissue sampling. The answer is that if tissue is inadequate for accurate diagnosis, it is even more inadequate as a basis for classification. If the diagnostic uncertainties are such as to preclude the identification of a particular tumor, it is preferable to leave it unclassified, until further data become available. To sum up, the creation of a separate tumor entity that is largely based on our inability to achieve diagnostic precision is counterproductive because, while it may provide to the less experienced a convenient name for tumors that cannot be diagnosed, it may result in the inclusion, in such a group, of tumors which either fall easily within one of the already established categories or which can be classified more accurately once a more detailed study of the case becomes possible.

Guiding principles for a classification of embryonal tumors of the central nervous system

How are we then to avoid the confusion I have described? It seems to me that we should perhaps be guided by two principles. The first is to adhere to the premise of systematic descriptions (or nosology), according to which distinct clinicopathologic entities are recognizable and therefore acceptable, even by those who have had little or no experience of them (as may be the case with the rarer entities). The old cartesian axiom still remains valid that, as much as possible, entities should remain clear and distinct, even though transitional or mixed examples may blur some of the distinctions. The cerebellar medulloblastoma is an example in point. It is a common tumor and has been extensively studied. Irrespective of its cytogenesis, it is distinctive because of three reasons: 1) the unique morphogenesis of the cerebellum, which is different from that of the forebrain, brainstem and spinal cord; 2) the mitotically active period of the germinative neuroepithelial precursor cells in that part of the central neuraxis extends from the greater part of fetal life up to the end of the first postnatal year, resulting in a long period of risk to neoplastic transformation and thus in a wide window of neoplastic vulnerability among tumors of central neuroepithelial origin (RUBINSTEIN 1987a); and 3) the relative frequency with which neuronal, glial, or bidirectional differentiation has been reported *in vivo* (RUBINSTEIN 1985b) and *in vitro* (HERMAN and RUBINSTEIN 1984; RUBINSTEIN and HERMAN 1986). All these features separate the cerebellar medulloblastoma from most other embryonal CNS tumors, in which, with the exception of two distinctive entities: the cerebral medulloepithelioma (SCHEITHAUER and RUBINSTEIN 1979; SATO et al. 1980; AUER and BECKER 1983) and the recently described desmoplastic infantile ganglioglioma (VANDENBERG et al. 1987a), divergent glial and neuronal differentiation is in my experience quite exceptional. In fact, save for the pineoblastoma, multipotential embryonal tumors arising in the supratentorial compartment and resembling cerebellar medulloblastoma still lack, despite contrary opinion (RORKE et al. 1985), sufficient documentation. Exceptionally, single cases have been described (AZZARELLI et al. 1977; DASTUR 1982; GAMBARELLI et al. 1982; TANG et al. 1985), but even here the details provided - and the documentation has then been adequate - make it clear that from the cytologic viewpoint those examples are different from typical cerebellar medulloblastomas.

With the exception of four examples which we subsequently identified as *desmoplastic infantile gangliogliomas* (VANDENBERG et al. 1987a), the tumors that we reported as cerebral neuroblastomas (HORTEN and RUBINSTEIN 1976; BENNETT and RUBINSTEIN 1984) show, when they do so (in about 25% of instances), differentiation only along ganglionic lines. The retinoblastoma is, *in situ* at least, very probably capable of differentiation only along neuroblastic or photosensory elements (PARENTES et al. 1987). The multipotential differentiating capacity that has been claimed to occur in the retinoblastoma cell lines Y79 and WERI-Rb1 (JIANG et al. 1984; KYRITSIS 1984; TSOKOS et al. 1986), if these observations are

TABLE I Current classical scheme of normal cytogenesis in the CNS relevant to embryonal tumors in Table II

Sites and stages of neurocytogenesis		Cell type	Differentiating potential
Forebrain	First	Primitive ventricular (matrix, neuroepithelial) cell	Multipotential (neuronal, glial and/or ependymal)
	Second	→ Neuroblast	Ganglion cells
		→ Glioblast ("spongioblast")	Astrocytes and oligodendrocytes
	Third	↓ Ependymoblast	Ependymocytes
Cerebellum	Neuroepithelial roof of the 4th ventricle	Undifferentiated neuroepithelial cell ("medulloblast")	Glia and neurons (?)
	↓ External granular layer	↓ Undifferentiated neuroepithelial cell	Internal granular neurons only (?)
Pineal anlage		Pineoblast	Pineocytes
Neuroepithelium of primitive optic cup		Retinoblast	Rod and cone cells

Arrows indicate directions of cytogenetic development

correct, does not seem to correspond to what actually occurs in retinoblastomas *in vivo*.

A second guiding principle should be that, irrespective of whether the classical scheme of normal central neurocytogenesis in the forebrain (Fujita 1963) may have to be modified in the future, a reasonably consistent approach to a classification of embryonal CNS tumors should preferably be within a framework in which their differentiating potential and the restrictions of that capability can be related to such a scheme. One should of course be aware that the entire concept of a histogenetic derivation of neoplasms is largely inferential and that in the case of embryonal tumors it rests on the number of assumptions for which the evidence may only be indirect. Nevertheless, a cytogenetic scheme of embryonal CNS tumors could provide a useful hypothesis to account for their differentiating potential. The current classic scheme of normal cytogenesis in the CNS, with the differentiating potential of the relevant cell types, is shown in Table I. Table II shows the embryonal CNS tumors which are the neoplastic equivalents of these cells, again with their differentiating

TABLE II Cytogenetic scheme of embryonal CNS tumors with distinct morphological features, correlated with Table I

Tumor entity	Presumed cell of origin	Differentiating potential of the tumor
Medulloepithelioma	Primitive ventricular (matrix) cell	Divergent {neuronal, glial, ependymal}
Neuroblastoma {cerebral, cerebellar*}	Neuroblast {cerebral, cerebellar}	Neuronal
Primitive polar spongioblastoma	Glioblast ("spongioblast") (?radial glia)	Astrocytic and/or oligodendroglial
Ependymoblastoma	Ependymoblast (?Ependymoglioblast)	Ependymal and primitive glial
Medulloblastoma	Undifferentiated bipotential cerebellar neuroepithelial cell	Divergent glial and/or neuronal*
Pineoblastoma	Pineoblast	Pineocytic, astrocytic, neuronal and/or photosensory
Retinoblastoma	Retinoblast	Neuroblastic and photosensory*

*Based partly on electron microscopy and/or tissue culture observations

potential. The entities also correspond to the progressive of differentiation that may be recognized in the neuroepithelial components of the OTT-6050 mouse transplantable teratoma (VANDENBERG et al. 1975; HERMAN et al. 1975) Those rare instances that do not fit into any of these diagnostic categories should in my view be left unclassified at this time. It has already been possible to separate from those unclassified examples an easily identifiable form which we have called *desmoplastic infantile ganglioglioma.* This is a distinct clinicopathologic entity in which the massive size of the growth is indicative of its pre- or perinatal origin and whose recognition is of biologic significance. Provided surgical resection is successful, it has a very favorable prognosis (VANDENBERG et al. 1987a). Its histogenesis remains speculative. It possibly arises from bipotential primitive neuroepithelial cells situated in the subpial granular layer of the cerebral hemispheres (BRUN 1965).

Contributions by modern techniques

The definition of the desmoplastic infantile ganglioglioma has largely been made possible because of the special neurohistologic and immunocytochemical techniques that established the divergent differentiating potential, glial and neuronal, of this distinct entity of infants. The systematic application of electron microscopy, tissue culture and immunohistochemistry is likely in the future to permit other very rare cases to emerge as separate morphologic entities, provided enough instances can be assembled and analyzed. As examples of entities that have been defined virtually solely with the electron microscope, we should quote the cerebellar neuroblastoma (SHIN et al. 1978; HIRANO and SHIN 1979; PEARL and TAKEI 1981; YAGASHITA et al. 1982; WARZOK et al. 1983; DE CHADAREVIAN et al. 1987) and the central neurocytoma of young adults (HASSOUN et al. 1982; PEARL et al. 1985; WILSON et al. 1985; OJEDA et al. 1986; TOWNSEND and SEAMAN 1986).

The argument has been advanced that those techniques are not useful for rapid diagnosis and therefore for treatment. In practice, however, the diagnostic and therapeutic problems that have to be confronted in this category of tumors concern only a few, very rare examples, since, as we have argued, the overwhelming majority can be included into an already defined entity once technically adequate material is available in sufficient amounts. Since the problems in many instances arise from the insufficiencies of a limited or improperly processed biopsy, adequate and careful sampling of the tumor by the neurosurgeon and, especially, satisfactory fixation of the tissue are, in this field, more essential than ever.

It is entirely proper that, as with other oncologic diagnostic problems, the resolution of these difficulties may have to be left to the more experienced neuropathologists. For it is precisely in those exceptional embryonal supratentorial tumors of infancy and childhood, in which divergent glial and neuronal differentiation can unequivocally be demonstrated, that there is the greatest need for documentation. If information on such cases is too sketchy or too general, then the different published series are no longer comparable, and the clinicopathologic correlations that may be derived become valueless. In instances in which either the sampling or the information is inadequate, or which for technical reasons cannot be sufficiently studied, the diagnostic formula "this tumor may eventually turn out to be" still retains its validity. The CNS is not the only tissue in which these diagnostic problems arise. Similar difficulties are currently being encountered with respect to the undifferentiated tumors of the peripheral nervous system (Dehner 1986).

Conclusions and future possibilities

The view presented here draws attention to some of the pitfalls of oversimplifying a difficult cytologic problem. Oversimplification limits the accuracy of diagnosis, may result in misinterpretation on the part of histopathologists less experienced than those forming today's discussion group, and may make it impossible to relate new series of cases to previously reported series or to biologic and cytologic information already available from the literature. The argument has been made that the preservation of distinct tumor entities, based partly on their localization within the central neuraxis and partly on their differentiating capacities, is more likely to constitute a useful basis for future advances, in contrast to the loose frame of reference that the simplistic concept of PNET's can provide (Vandenberg et al. 1987b).

Advances in cell type identification by special neurohistologic, immunohisto- and immunocytochemical techniques have permitted the discrimination of distinct cytomorphogenetic entities. These are based in part on their resemblance to the normal phases of neurocytogenesis. Four of these tumors, the cerebral medulloepithelioma, the desmoplastic infantile ganglioglioma, the pineoblastoma and the cerebellar medulloblastoma, are pluripotential in their capacity to exhibit divergent differentiation. Cytomorphogenetic, clinical and experimental data implicate fetal neural cell targets for transformation and raise the possibility that aberrant developmental regulatory mechanisms may contribute to the biologic behavior of these tumors. Growth factors and neurotransmitters (the latter of which are known to act as modulators of normal neuromorphogenesis) could play a regulatory role in central neuroepithelial tumors: the hypothesis is that the aberrant behavior of embryonal neoplasms could be modified by functional receptor responses or result from abnormal receptor responses to those substances. Future possibilities in this complex field include: 1) the definition of new cytomorphogenetic entities and subgroups of the currently defined forms of embryonal CNS tumors, perhaps based on the presence of specific growth factors and regulatory neurotransmitters, or their receptors, 2) the characterization of receptor responses mediating the modulatory role of growth factors or regulatory neurotransmitters on the growth and maturation potential of these tumors and 3) the definition of developmental, stage-specific modular that might be operative in this group of neoplasms (VANDENBERG et al. 1987b).

Discussion of Chapters I and II

DR. ARMSTRONG: We welcome all of you to Houston and we hope that you will find this a stimulating and exciting meeting. We recognize that you are here because you are concerned as we are with the naming of things. Shakespeare felt that names were not important and that a rose by any other name would smell as sweet. But we know how important names are in addressing our colleagues, in surgery and oncology, therapeutic radiology, and now in basic science. So we hope to go forward with this meeting in the spirit of finding proper and meaningful names.

Floyd GILLES (1983) reviewed the problems of naming, and I would refer to that very clearly written article in which he described the ways in which tumors are now named. He said that tumors are named because of what we think they look like, or because of where we think they came from, or because of where we think they are going. He also said that it is probably very important in choosing a name to decide whether we are going to talk about the cell of origin, or the etiology, or the prognosis. He suggested that it may be impossible, in one name, to fulfill all of these objectives.

"You don't look like a Canadian."

Drawing by Drucker; circa 1967 - The New Yorker Magazine, Inc.

The preceding figure illustrates the problem with employing an all-inclusive name such as *Canadian* when we really do not understand where the Canadian came from, whether he had French, English, Chinese, or Yugoslavian origin. We do not know why he is a Canadian or how he will behave in the free market, or at the Olympics. Nevertheless, we are very much attached to that name *Canadian.* It has emotional significance to many, and if we negate the name, we negate history. If we negate history, we negate the present, and we have no future. Therefore, it is very important that we remember the place of history when we are deciding about the correct name.

Today, we think about the history of the nomenclature of brain tumors. We are reminded that VIRCHOW (1864) was the first person to identify that glial cells were different from neurons, and that RIBBERT (1910) was able to identify, or thought he recognized a similarity between tumors and the way the nervous system develops. Since then, there have been many attempts to name tumors, in relation to the way we think the nervous system develops and the way we think tumors resemble those embryonic cells.

We are at a most exciting time in the history of histology. We have, in addition to our great heritage of the classic techiques of neuropathology, the challenging opportunity to apply the new techniques of molecular biology. Today we can, with antibodies, identify the surface of cells, we can look into the cells, we can identify their genes, we can identify gene amplification, and we can identify gene products.

We hope that at this meeting we will think about the utilization of these new techniques in naming tumors so that we can better serve our patients with more directed therapy, perhaps initiated by the name that we give to the tumor.

With both the past and the future in mind, our discussions will inevitably lead us to dealing with the concept of primitive neuroectodermal tumors. This name was introduced by HART and EARLE (1973) in their description of 23 tumors of the nervous system in children, which they felt did not quite fit into any specific category of tumor. One year earlier, RUBINSTEIN (1972) used the term *primitive central neuroepithelial tumors* of the brain in relationship to a concept of histogenesis and cytogenesis of cells in the nervous system.

In 1983, Lucy RORKE introduced her concept of primitive neuroectodermal tumors. The following year, a group of pediatric oncologists in North America met at Niagara-on-the-Lake to discuss the needs of physicians dealing with pediatric brain tumors in relationship to brain tumor classification. They examined the present WHO classification and decided that there were areas in which the descriptive requirements for the pediatric group of tumors were not met.

Dr. Larry Becker reviewed the WHO classification. Dr. Peter Burger examined the way in which a classification that better defined histologic variants could possibly be formed in the future. Dr. Floyd Gilles suggested alternate ways of classification, which he will discuss later, and Dr. Rorke presented her proposed revised classification for childhood CN8 Tumors.

Today, we have been privileged to hear Dr. Rorke define for us her concept of the primitive neuroectodermal tumor and then to hear Dr. Rubinstein present his rebuttal.

DR. BARNARD: We have heard two extremely interesting papers that have presented clearly two different aspects of a problem with which we are very familiar. Essentially, the difficulties that we have are compounded in two ways.

One is that neuropathology is now becoming increasingly a service directed to the investigation of small biopsied material and we no longer have at our disposal the postmortem studies that we would like to have in order to elucidate problems. Therefore, in our classifications we must try to find ways for helping people who may not always have enough material to investigate things properly.

Secondly, we have the problem that whatever we decide has to have a worldwide application. We are basically trying to provide a classification for people who may be in the field of general pathology and in circumstances where they may not have very experienced neuropathologic diagnostic skills available.

In this particular field, and I am dealing here with the concept of the embryonal and the primitive tumors, two points of view have been put forward. In her paper, Dr. Rorke very elegantly expounded the view which we have all read in her publications, that there is an advantage in putting forward a concept of a primitive neuroectodermal cell that is not confined to a particular site.

This concept was first put forward by HART and EARLE (1973) to establish a category for those difficult undifferentiated cerebral neoplasms which do not have obvious names because it is often very difficult to define them. Subsequently, this has been extended to include several familiar entities, such as the cerebellar medulloblastoma.

Then we had Dr. Rubinstein pointing out the importance of retaining the many well-established, well-understood features of familiar tumors such as medulloblastoma, and warning of the dangers in applying overall terms both to defined entities and to ill-defined entities. He also made it very clear that we have a particular problem with respect to the question of differentiating anaplasia from embryonal characteristics, to put it in that kind of way.

DR. ISHIDA: The comprehensive term, *PNET*, for undifferentiated childhood tumors, has received increasing attention among Japanese neurosurgeons because this designation is quite simple, covering all types of embryonal CNS tumors. It is also of some practical importance because neurosurgeons can predict the expected clinical behavior of a given tumor. However, by using this generic term I am afraid that the characteristic features of an individual tumor become faded. Clinical behavior and cytology of embryonal CNS tumors appear somewhat different according to the site from which the tumor originates. Pineal cell tumors and cerebral and cerebellar neuroblastomas are such examples.

BAILEY and CUSHING (1925) described four cerebral examples of medulloblastoma in their original series of cases. I would prefer to confine the term *PNET* to the embryonal CNS tumors of the cerebrum in children.

DR. BRUCHER: I just have one question. Dr. Rubinstein said that there is a difference between anaplastic tumors and embryonal tumors. Perhaps it is necessary for us to have a definition of anaplasia because embryonal tumors contain small undifferentiated cells. What do you mean when you say that there is a difference between anaplastic and embryonal tumors? I think it is a very important point.

DR. RUBINSTEIN: This is an example that justifies Dr. Sobin's distinction between a conceptual and a descriptive approach to the problem.

In terms of what one sees, there may be no morphologic differences between anaplastic and embryonal tumors. An example well known to us is the cerebellar medulloblastoma which shows astrocytic differentiation. These tumors have been reported under two different designations. Some have regarded them as astrocytomas showing anaplastic change, culminating in the picture of medulloblastoma. The same type of tumor has been reported by others as medulloblastoma showing astrocytic differentiation.

The difference lies in our concept of what has happened. In practice, when I receive such a case, my response is: "It may be a medulloblastoma differentiating into an astrocytoma, or, alternatively, an astrocytoma becoming anaplastic. So far as treatment is concerned, treat as for a medulloblastoma." Thus, I recognize the distinction between what I actually see and the way I conceive what I am seeing. It may, however, be that the distinction is artificial, since it is probable that most brain tumors arise from the neoplastic transformation of a cell in the course of CNS development. The more extreme view then would be that, conceptually, all gliomas are PNETS.

DR. KLEIHUES: I would like first to make a more general statement. Ever since the publication of the paper by Dr. RORKE (1983) in which she first described this concept, I was very fascinated by her idea. I think it, more than many other papers, has provided an understanding of new concepts regarding childhood brain tumors. It has had a very great impact on neuropathologists in Europe and has opened many discussions with neurosurgeons and radiotherapists who previously were not aware of such a hypothesis.

I believe hers was a very important paper that really opened a new door. Nevertheless, I must say that although I liked it very much as a concept, what has been made out of the proposed classification is something with which I cannot fully agree. The main reason, in my view, is that the original very attractive hypothesis has been misinterpreted.

I realize that in some cases it may be almost impossible, with the methods available to us at the present time, to decide whether a tumor is anaplastic or is derived from a precursor cell population not yet undifferentiated. However, in my opinion, this problem is not very frequently encountered. So for this purpose, to include tumors which you are not sure are anaplastic or coming from a precursor cell population, leads you to a certain ambiguity. If you call all these tumors *PNET*, then, of course, you will be tempted to call all or practically all malignant childhood tumors by this name.

I must agree with Dr. Rubinstein that increasingly one receives papers submitted for publication, including papers from radiotherapy journals, which put all of these tumors together. In my view, this has resulted in an untimely practical result which has distorted a very valuable concept. I think development of this concept, which is actually attractive as a concept, must be put into a workable and useful classification for routine work.

I would like to add that I am of the opinion that one of the bases of the concept that Dr. Rorke developed was that these tumors, irrespective of their site of origin, have a similar biologic behavior. I do not entirely agree with this assumption. I believe that we will increasingly recognize that, although they may all look like medulloblastoma at one stage, they actually have different potential and are not exactly the same. I am sure you will soon find that they do not behave exactly the same with certain types of treatment.

In particular, I think at this meeting additional evidence will be proposed which shows that on the basis of tumor genetics, we must more clearly define new concepts which may not readily be understandable to the morphologist. If you keep in mind that the peripheral neuroepithelioma and neuroblastoma can at present virtually be distinguished only on the basis of genetics, this may very well be true for many tumors which some of us now call *primitive neuroectodermal tumors*.

DR. KEPES: I would like to comment on the question raised by Dr. Brucher about the meaning of anaplasia as compared or contrasted with the primitive state of the cell. I think, and please correct me if I am wrong, that there is both a narrower and perhaps a wider meaning to the term *anaplasia.* In the narrow sense, it simply means that the cell that used to be differentiated is now returning to a more primitive state, to wit, the mucous-producing lining cell of the colon which is now reverting to a more primitive, non-mucous-producing type, or the keratin at the site in the epidermis which is now becoming like a basal cell that does not produce keratin. So they meet halfway, the primitive cell and the anaplastic cell that has reverted to a more primitive state. This is certainly one feature of neoplastic cells.

However, in my mind it is important to remember that in everyday practice, the term *anaplasia* means more than that. Some use the term *anaplasia* for other types of cytologic malignancy, such as bizarre multinucleated giant cells, polypoidia, mitoses that are tripolar. You do not find tripolar mitosis in the normal developing cells of the central nervous system, or of the gut. Therefore I think we have to make clear in our minds what kind of anaplasia we are discussing.

DR. BURGER: I was present at the deliberations during the conference in Niagara-on-the-Lake, and I am the chief neuropathology reviewer for the Pediatric Oncology Group, so I have had a chance to see the evolution of the *PNET* concept and its effect on clinicians and pathologists. At the meeting at Niagara-on-the-Lake, when it was presented by Dr. Rorke, there was considerable acclaim for it as a new understanding of these childhood tumors by the clinicians who were present. As a consequence, I, as a pathologist in the Pediatric Oncology Group, was within the next several months expected to classify all medulloblastomas as *PNETs*. I resisted this since we will know nothing more about medulloblastomas because we call them *PNET*s, and encourage the false concept that medulloblastomas are biologically equivalent to supratentorial PNETs.

I have also resisted using the term *PNET* for many of the supratentorial lesions. We frequently receive pieces of tissue for histologic review that are poorly processed or very limited in size. In such situations the clinicians are often not satisfied with a diagnosis of an undifferentiated tumor. They would be much happier with the diagnosis of a PNET. I have resisted using the designation *PNET* in such cases since it would create the appearance that we understand an entity when we do not.

The other consequence of the PNET system was on pathologists. Many of them adopted this system, partially because it is easy. There is no question that these primitive tumors in children are difficult to classify and some generic term would make this task easier. It is very easy to diagnose a PNET or any neoplasm with a component of undifferentiated cells. Because of this, this term has become overused.

Another perspective is that of the radiotherapist. They are generally not in a position to inquire into the details of these classifications, and there is quite a tendency for them to lump neoplasms into groups with the assumption that they are all equivalent. I think it is clear that the medulloblastoma in a significant percent of cases, perhaps half, is cured by radiotherapy. It would appear that the majority of supratentorial PNETs are not. This to me is cause for quite some concern if both classes of tumors are called *PNETs*.

So in summary, I think the PNET concept is an excellent idea, a concept, and I think undoubtedly some medulloblastomas are true PNETs. However, I suggest that we retain the term *medulloblastoma*, in light of its historical recognition as an entity, and be extremely cautious about using the term *PNET* for supratentorial lesions.

DR. SCHEITHAUER: My concern is the same as that of Peter Burger. He and I have had many discussions about the PNET issue. I think a part of the difficulty is that the concept, due to its simplicity, is very enticing. Where I have a problem with it is in its practical application. We have an extensive therapeutic literature, perhaps more vast than that of pathology, on the subject of medulloblastoma and other embryonal tumors.

To suddenly alter terminology has its costs. It is important to recognize that at present the scheme of the embryonal neoplasms available to us is workable in the majority of instances. In exceptional cases, the difficult-to-pigeonhole cases, a problem of classification arises. More often than not it is merely a problem of tissue size or sampling rather than anything else.

The difficulty that we are going to face if we adopt a classification that permits the use of a broad non-specific term, is that the users of the classification, most of them practicing general pathologists, are going to abuse the term. Many do not have immunocytology, and some do not even have an adequate PTAH stain available to them. These physicians are going to be handicapped in their effort to subclassify embryonal tumors.

The overwhelming majority will make a diagnosis of *PNET*, and their cases will not be further studied. The problem is not just failure to properly classify small cell embryonal tumors. I have seen dedifferentiated gliomas, anaplastic oligodendrogliomas, lymphomas, small cell carcinomas, and even undifferentiated sarcomas, classified as *PNETs*.

When all of this lumping has been done, how will we be able to compare publications, based on diagnostic homogenization, emanating from a variety of institutions? I think that inter-institutional comparison of the results of therapeutic protocols is something we should be striving for. Historically, even in our classification of astrocytic tumors, there have been problems of recognition and grading. In my estimation, these small-cell neoplasms are more difficult to classify. If we change the designations, the benefits had better be overwhelming in some way. In this instance, I don't believe that they are.

I agree with Prof. Kleihues, that now is the time new methodology is permitting us to split, sub-split and better characterize neoplasms. We want very much to arrive at understandable, reproducible diagnoses. The embryonal tumors do not need a simple terminologic shakeup. In my own institution, I have occasionally used the term *PNET*. I reserve the term for small cell lesions I cannot classify honestly, not just because it is a poor specimen, but after thorough workup. If we cannot identify it or if it shows so many varying components that there is simply no place in the present classification to put it, we might then have a legitimate use for the term *PNET*.

Alternatively, if the specimen is very poor, or there is not enough tissue to go by, perhaps the intellectually honest thing to say is that this is simply a malignant small cell neuroepithelial tumor, type otherwise not classified.

DR. DAVIS: I think we have actually at least two roles in this, one as scientists, and the other as teacher-educators. As scientists I think we need to keep looking at the tumors with all the techniques available to us, in an attempt to find out if, for instance, differentiations in the small cell tumor occurring in the posterior fossa give it a better or a worse prognosis. I think we need to look at that.

The other thing we need to do is to educate our colleagues, both neurosurgical and neuroradiologic colleagues. I think we can do that by having a transition period (which I believe was a part of the original proposal) in which we diagnose, for example, medulloblastoma as a *PNET* occurring in the cerebellum with or without one kind of differentiation, and also use *medulloblastoma* as a subdiagnosis until people are educated to the new terminology.

DR. KLEIHUES: What do you call them?

DR. DAVIS: I call them both.

DR. WECHSLER: Well, I think the main purpose of this session was not to convince us that this is an exact concept, or that the cytogenetic concept of tumor classification is the right one. It is merely a clear demonstration that we have to continue our studies and look for modern basic research tools in this very complex field of neuro-oncology.

As far as the accepted WHO classification (1979) is concerned, I think neither we nor other colleagues in many countries have great problems with the diagnosis of a medulloblastoma of the cerebellum on an H & E section. As far as tumor grading is concerned, it was proposed by the WHO classification that all medulloblastomas should be considered as grade IV. Therefore, tumor grading of medulloblastomas has not become a special topic in neuropathology.

Immunohistochemistry offers very important markers here, e.g. the proliferation associated antigen, Ki-67, originally described by GERDES et al. (1983). With this method it is possible to evaluate the proliferation potential or, in other words, the degree of biological malignancy of an individual human brain tumor in frozen sections at the time of operation. In the few publications on the subject, BURGER et al. (1986) studied 40 tumors but no medulloblastomas, and GIANGASPERO et al. (1987) determined the growth fraction in 22 primary brain tumors, including 3 medulloblastomas. The percentage of Ki-67 positive cells in these medulloblastomas was between 40-50% with a mean of 43% and a relatively narrow standard deviation of 5.7. In our own material we had the opportunity to study 9 medulloblastomas in which the range was between 5-42%, representing a mean of approximately 18% and a standard deviation of 13.5. Therefore, I think we should not only concern ourselves with the very important problems of cytogenesis and cytodifferentiation of this primitive neuroectodermal tumor type, but also with the growth behavior by using these new techniques. They offer a better understanding of the proliferation potential, indicating that a medulloblastoma may grow either more rapidly or more slowly with measurements of proliferation indices between 5% and 50%.

DR. HASSOUN: I wish only to express some suggestions and make a few observations. First, with respect to the problem of differentiation, I think it is very difficult to be precise about the differentiation of tumors by light microscopy alone. This kind of statement depends on the technical approach. The level of differentiation would be differently appreciated by light microscopy than by electron microscopy, or by immunohistochemistry. We have to keep this in mind as a very important point.

For instance, 210 kD neurofilament immunoreactivity is often absent in some genuine neural tumors. That is not to say that we are not dealing with neural tumors. There are many other neuronal markers and we must look for as many of them as possible. Of course, for the small biopsy specimen that we usually have to study, electron microscopy appears to be a very useful and reliable tool for diagnosis.

Concerning the problem of histogenesis, I do not see why we have to choose between a process of anaplasia and an embryonic process. Both of them are possible. Embryonal tumors are more frequent in children, and the process of anaplasia is probably more frequent in adults.

One further observation concerns mixed tumors. Why wouldn't it be possible that the neoplastic process bears both on glial cells and other cell types? For instance, it is obvious that in glioblastomas, the neoplastic transformation involves astrocytes as well as blood vessel cells. It is certainly possible to find these multifactorial neoplastic processes in other kinds of brain tumors such as sarcomas, vascular tumors, etc.

DR. ARMSTRONG: Dr. Rorke, would you like to respond to these challenging questions?

DR. RORKE: Well, it is interesting for me to hear all of the comments, and I am pleased to have the chance to respond. First of all, I would like to respond to Professor Kleihues' discussion of both papers already presented.

As so often happens when some controversial idea is cast into the community, a variable response is elicited. When I was invited to organize a committee to examine the problem of the classification of childhood brain tumors, I told the person who called me, that I did not think this would be possible. Neuropathologists have a difficult time agreeing on anything. Moreover, I knew that the problem which was giving the neuro-oncologists difficulty was what should be done to clarify definitions of this category of tumors with a population of primitive neuroepithelial cells.

So after long discussions, relative to the problem, during which my colleague was rather insistent, I finally said, "Alright, we'll see what we can do." The members of my committee were neuropathologists with long experience in diagnosing childhood brain tumors, some of which stemmed from the study that Dr. Floyd Gilles had organized to review 3,300 childhood brain tumors. Dr. Richard Davis, Dr. Larry Becker, Dr. Gilles, and I were all involved in that project. Specifically, Dr. Davis and I looked at all 3,300 childhood brain tumors. Our assignment was to give a WHO diagnosis to each one of these tumors.

When I was asked to organize this committee and to develop a classification for the childhood tumors, I had grave misgivings. Since we had already used the WHO classification for the 3,300 tumors and had found it to be quite valuable for diagnosing many of them, we thought it prudent to start with that, and then examine those categories for which the fit was not so apparent.

We had relatively little difficulty until we came to this group of embryonal tumors. In fact, we spent the majority of time discussing this group of tumors. What was finally published, you should understand, was a compromise solution to the problem. Now, perhaps compromise is not an ideal way of arriving at conclusions about such matters. After our exhaustive (and exhausting) discussions, we agreed to have the word *primitive* surrounded by quotation marks. That was done to direct attention to the fact that this issue remained controversial.

The other aspect of the problem with which we had so much difficulty was the use of the word *differentiation*. We finally decided that we, as pathologists looking at the slides, really could not determine whether the cells had actually originated from the primitive state along one line or the other, or whether they were transformed cells. For that reason we simply utilized the word *astrocytes* or *ependymal cells* or whatever we thought the cell type was.

I am sorry if you are disappointed that this particular proposal is somewhat different from my initial presentation in 1982. Under those circumstances, the entire responsibility was mine and I could say what I wished in my presidential address. My object in raising this issue was not to disagree with those who had done serious work, which I respected, but rather to call attention to some of the inconsistencies in the classification schemes, which I thought did not fit into a logical scheme. One of the things that impressed me during this review of the 3,300 childhood brain tumors was that when Dr. Davis and I looked at the slides, we had no idea what the origin of the tumor was, e.g., we did not know where in the nervous system the tumor was located. We could not, therefore, make a diagnosis of medulloblastoma or cerebral neuroblastoma, or pinealoblastoma, by simply looking at the slides. It may be perfectly acceptable to use the term *pinealoblastoma*, for example, when we know that the surgical specimen came from the pineal area. On the other hand, if we do not know its source and are going to be intellectually honest, we cannot really say whether the tumor is a pinealoblastoma or a medulloblastoma or one of the other types of primitive neuroectodermal tumors.

Moreover, if we check descriptions in the standard textbooks on histologic criteria that would allow separation of a medulloblastoma that contains no other cell type, from a tumor that looks similar but arises in the pineal gland, we cannot do it objectively.

It was for these reasons that I thought that the classification system in use was inadequate and was therefore emboldened to propose a change in the terminology. It is perhaps simplistic, but I guess I have a simplistic brain, and I can't think along complicated lines. Furthermore, I felt it was a great deal more logical, to do it that way.

DR. JELLINGER: The term *PNET* appears similar, at least for general pathologists, to sarcoma, for example, or even to each small-cell mesenchymal tumor wherever it arises. I think we should come to a compromise between these two suggested proposals.

The term *PNET* covers a large group comprising a number of different types of tumor, perhaps of the same origin or not, which might be proven by more modern genetic methods As in general pathology, it is not possible to make a definitive diagnosis only from H&E preparations. We have to also use still more sophisticated methods.

It is of little use, particularly for the clinician as well as for the radiotherapist, to use one term, sarcoma or PNET, for a rather primitive tumor, which needs to be classified by more sophisticated methods. Moreover, routine diagnosis is not enough to prove that this is a primitive small-cell tumor, whatever it may be. We are obliged to use all the techniques available, along with routine diagnostic methods.

DR. RUBINSTEIN: Everyone has made cogent points and all the points are highly relevant to our discussion indicating that in many aspects we are in agreement. Nevertheless, the individual way of looking at these problems must to some extent be respected. Where there is no objection to people using any terminology, they favor in their own laboratory or institution, I have an instinct against imposing in a prescriptive manner a specific terminology with the claim that a consensus has been achieved.

As you know, our meeting is partly the result of the Niagara-on-the-Lake meeting which was chaired by Dr. Rorke. I would like to say two things.

First, I am not convinced that two individuals looking together with a doubleheaded microscope at small pieces of tumor without knowing where the tissue arose can perform a useful exercise in tumor classification. An oligodendroglioma may look just like a peripheral neuroblastoma. Some osteogenic sarcomas may resemble anaplastic giant-cell astrocytomas. One should never attempt a diagnosis without knowing where the tissue comes from, without knowing the age of the patient, the sex, the exact localization of the specimen, etc. In practice, we just do not function in this manner, and the localization of the tumor is just as important within the central nervous system as outside it.

The second point is also a practical one which should be repeated: the terminology introduced a few years ago by Dr. Rorke and her colleagues has led to such a problem of misdiagnosis on the part of nonspecialists that it cannot possibly serve as a firm basis for future classification. This is not to deny, of course, that I agree with some of the points made by Dr. Rorke.

DR. ARMSTRONG: There was reference to the problem of an undifferentiated or anaplastic astrocytoma versus a medulloblastoma of the cerebellum. Dr. Rubinstein, I believe, said that it should be treated as a medulloblastoma. In your laboratory, would you treat a medulloblastoma and an anaplastic astrocytoma the same way?

DR. RUBINSTEIN: There are tumors in the cerebellum, particularly in children, which look like medulloblastomas and yet show astrocytic differentiation. There are also astrocytomas in which the anaplastic areas are indistinguishable from a medulloblastoma. My perception is that the appearances are interchangeable, but that as far as treatment is concerned, one would treat as for medulloblastoma. I am, however, differentiating such tumors from those rare anaplastic astrocytomas or glioblastomas that may occur in the cerebellum of adults. These present a different problem.

DR. KLEIHUES: But if you do not know, then you treat them as medulloblastoma?

DR. RUBINSTEIN: Yes.

DR. ARMSTRONG: Dr. Scheithauer, you referred to some problems with the differentiation of the PNET versus lymphomas or malignant oligodendrogliomas or metastatic sarcomas. Were these in children or is that more of a problem in adults?

DR. SCHEITHAUER: Not all, but several were in children. The undifferentiated sarcoma was in a child, whereas the lymphomas and anaplastic gliomas were in adults. This is the problem. The diagnosis of PNET is being used far too frequently in adults. I think that is one of the problems with the concept. The same difficulty with overuse of the term is also encountered outside the CNS with small-cell neoplasms in lymph nodes, and other organs, such as the nose. Some of the neuroepithelial tumors at this site really look like neuroblastoma, whereas others look like paraganglioma replete with sustentacular cells and still others look like carcinoid tumors with gland formations. Lastly, some nasal small-cell tumors are basically undifferentiated carcinoma, that look like oat cell carcinoma of the lung. At present all of these can be classified separately because we have established criteria for their distinction. But if we use the term, or equate the term *neuroblastoma* with PNET, and continue to enlarge our concept of neuroectodermal tumors of the nose, all of these relatively undifferentiated lesions will soon be considered simply PNET.

DR. KLEIHUES: You asked the question whether and to what extent we are actually using the term *PNET* in clinical diagnosis. We do use it and we stick to your original concept, strictly. That means that we use it for a tumor which looks like a medulloblastoma but is not located in the cerebellum or in the pineal gland, it is located somewhere else. If we have employed all the methods available to determine that it is not anaplastic glioma and we are convinced it is a tumor arising from the precursor cell population and it is not in the cerebellum, not in the pineal gland, then we call it *PNET*. If we are not sure whether it is an anaplastic glioma or such a tumor, then, we merely say this is an undifferentiated small cell tumor.

How do we exclude the possibility that it is an anaplastic glioma or whatever? It was my intention to talk about this later in more detail, but a few remarks might be appropriate at this point.

We have been fortunate over the past year to look at 230 medulloblastomas from the International Science Study. We have looked at glial differentiation in all of these tumors, using all the appropriate markers available. In our view, extensive glial differentiation of these is very, very unusual. You can find distinct neoplastic cells expressing GFAP in less than 10% of the cases, and we have never seen in these 230 cases real areas of progressive astrocytic differentiation.

When we look at tumors of this kind, we look for GFAP, and if we have no GFAP expression whatsoever, or if we have the kind of expression which we have seen in about 10% of medulloblastomas, then we assume that this is indeed a primitive neuroectodermal tumor. Occasionally that may not exclude the possibility that it is an anaplastic glioma, but if we have the feeling that we cannot exclude it, then we do not put it down as *PNET*. Otherwise, we would be calling it a *primitive neuroectodermal tumor* when we do not know where it originated.

So we take a kind of a practical approach, but we stick to the original concept that these are tumors which derive from a precursor cell population, and they are indistinguishable from medulloblastomas. I know that in many cases they are distinguishable. There are many pinealoblastomas in which we can easily distinguish, for instance, the pineocytic differentiation, which does not occur in the cerebellum. Some cases are easy to classify but there are others which we cannot distinguish, although they look like medulloblastomas. If there is no indication that they are anaplastic gliomas, I think for these cases the term *PNET* is appropriate.

DR. BIGNER: I was going to wait until later when I discuss some experimental data, to make this comment, but I think there is a general problem, that should come up in this discussion now. I am very concerned about pathology as a discipline right now. We are under attack as a discipline in the U.S. And usually, when that sort of thing begins here, many of you in Europe and Asia and other parts of the world will see the snowball developing as well. I believe these attacks are justified in part, because many pathologists are clinging to old concepts and methods, and are not introducing new molecular methods and concepts fast enough.

Whether we like it or not, more rapid progress is being made in tumor biology now than in the last hundred years. And we are not going to solve these problems of tumor classification adeguately with more than one-hundred-year-old technology such as light microscopy and hematoxylin and eosin stains. What I see as a fundamental problem in reviewing manuscripts on these things, and discussing data is an inadeguacy bordering on technical incompetency, of most pathologists to conceptualize and execute in a critical way, these newer classification methods. We are not going to be able to continue to simply use light microscopy other than our fundamental basis for initial communication about classification.

This new classification and new biology is going to come whether we like it or not, and I would like to see us, as pathologists, stay a part of it and be in the forefront of applying the new biology. It would be much better, rather than simply having to come along after the work is done by neurobiologists, molecular biologists, or general cancer specialists, that we be part of new molecularly-based tumor classification schemes.

To make an even more specific point, most of the immunohistochemistry that is done in trying to answer these questions is performed inadequately from a technical standpoint. One cannot take most of the new monoclonal antibodies and work retrospectively with paraffin embedded tissue. There are three principal problems in using monoclonal antibodies, and most neuropathologists and general pathologists do not understand these three principles.

First, the majority of monoclonal antibodies will only react with epitopes that are present in living or frozen tissue. Formalin, alcohol, glutaraldehyde and other common fixation methods, followed by routine paraffin embedding and processing, destroy or markedly diminish monoclonal antibody reactivity by destroying or changing epitopes.

Second, with many of the intermediate filament proteins, many enzymes and other molecules, the most important problem is the length of time that tissue has been out of the body before freezing or fixation. If tissue has laid on the surgeon's Mayo stand and sits there for an hour or two hours before it goes down to pathology for processing for histology, autolysis and degradation of many of these proteins and other molecules occur and one is not going to detect the epitopes even if tissue is frozen and the appropriate monoclonal antibody is used.

Third, many monoclonal antibodies, for example A2B5, are not monospecific or epitope defined. Such antibodies as A2B5 may react with many different cell types or molecules, and be very difficult to interpret in use with neoplastic cells. Finally, even epitope-defined or monospecific monoclonals may react with very different types of molecules, for example with carbohydrate epitopes shared by highly dissimilar glycolipids and glycoproteins.

So I take a very jaundiced view of these retrospective analyses with the handful of available markers in paraffin embedded tissue. I do not think we are going initially to be able to use such data to determine with any degree of accuracy what markers are expressed or not. The required process, which is likely to be laborious and time consuming, will be to take, for example, some of the new types of antibodies (UJ181.4,UJ13A) that John Kemshead has developed in London (KEMSHEAD et al. 1983), that are not even characterized in terms of their reactive molecules, that will only react with live cells or with frozen tissue and perform immunohistochemical analyses.

Only when new tumor-associated molecules have been purified and there are second generation monoclonal or polyclonal, i.e. monospecific polyclonal antisera, that then can be used retrospectively and prospectively on paraffin embedded tissue, can large surveys on routinely processed tissue be reliable.

Moreover, we are going to have to be very, very careful about the broad interpretations that we make about these markers being present or not except in the types of analyses which I have just described.

DR. RORKE: I think there is some confusion as to what my original concept was, and I would really like to restate it to clarify in your minds what I was trying to get at when I introduced this concept of primitive neuroectodermal tumors. I specifically excluded the olfactory neuroblastomas neuroectodermal tumors. I specifically excluded the olfactory neuroblastomas and the retinoblastomas because I felt that these tumors had been clearly delineated. They were in a unique location and they really did not fit in with the problem that I perceived was occurring with the group in the brain and spinal cord.

Basically, the concept that I introduced involved what I thought was the equivalent of the peripheral small round-cell tumors in the nervous system. I thought that neuropathologists at least were sufficiently well trained that they would not confuse an oligodendroglioma with that category of small roundcell tumor. Therefore, I did not even take that into consideration.

What I did was borrow a concept from the general pediatric pathologists, who recognize a group of small round-cell tumors in other organs of the body, and who found that when they applied marker technology to the investigation of these small round-cell tumors, they could identify certain subcategories of tumors, namely rhabdomyosarcomas, neuroblastomas and so forth.

Thus, when I introduced this concept for the central nervous system tumors, I took the group as a whole regardless of where they occurred. These were parenchymal tumors that consisted of a population of small round cells and were otherwise not distinguishable by routine light microscopy. It was these tumors that I put into the category of the primitive neuroectodermal tumor. I then suggested that we could separate them by looking at them in the same way that the pediatric surgical pathologist had done, i.e. by use of various marker studies to see whether they were differentiated along one line or another, by utilizing electron microscopy, or any other technique that would identify the cell population that made up the tumor. I did not conceive of this as a wastebasket category as I thought had been represented by the introduction of the term by Hart and Earle (1973). I did include under this umbrella, of course, the medulloblastomas because they comprise the largest number of such tumors in the nervous system.

I am distressed by the extension of the term by others to all include the peripheral examples. However, I cannot be responsible for the thoughts of others nor should I be held responsible for other people's laziness or wrong diagnoses. We should separate those who take the easy way and use the term for all tumors presenting difficulty in diagnosis from the original intent of what the concept was meant to be. In other words, I was attempting to clarify some of the confusion relative to terminology of these tumors and the manner in which they were defined.

I certainly decry utilization of the blanket term *primitive neuroectodermal tumor* for every tumor that contains small cells and that we cannot otherwise identify. That is a total misuse of the term and reflects a lack of understanding of what I initially proposed.

DR. DAVIS: I do not think we need to be responsible for another person's misdiagnoses, I just do not think that is my job. I try to teach people, who do not see as many of these tumors as I do, about them. I think it is healthy that our colleagues, who are not so specialized, send their slides or tissue in for our opinion and take advantage of our greater experience.

DR. KLEIHUES: I do not entirely agree with you. If you propose a classification and the result is that people start to confuse things which should not be confused, you might not be responsible for the wrong diagnosis, but you are still responsible for the classification which perhaps created the confusion.

DR. RUBINSTEIN: There is one point I am forced to make. I think there was little or no confusion before the introduction of the term *PNET*. All of the confusion seems to have arisen after its introduction, so I think this is where the confusion lies. Not before, but after.

DR. DAVIS: They just were not sending in the cases then.

DR. RUBINSTEIN: Oh, yes, they were.

DR. ARMSTRONG: Could I ask Dr. Burger to comment on the remark that he made earlier? He mentioned that he uses the diagnostic term *medulloblastoma* and that half of the patients do well and half of them do not. How can we separate the ones that do well from the ones who do not?

DR. BURGER: You cannot. There are three studies that I am aware of. One claims that cellular differentiation is a bad prognostic sign; another suggested that differentiation is a very good prognostic sign; and still another study in adults suggested that none of the factors have any prognostic significance. I have a hard time believing there are not some histologic factors of diagnostic significance in the medulloblastoma but this has not been established conclusively. I think the cooperative study for which Professor Kleihues reviews the pathology might answer that question. It is scheduled to encompass 300 or more patients, all studied in quite some detail.

I think there is yet to be a published study that delineates in a large number of cases, how many were neuroblastic, astrocytic, and undifferentiated, etc. and what the survival figures were for patients with each type of neoplasm. It would also be important to illustrate the criteria used.

DR. FIELDS: If a stranger were to walk in here in the midst of this discussion, and try to decide exactly where we are going with all of this, he would have a terrible time. I have heard that 80 or 90 years ago neurosurgeons were very unhappy with the pathologists because they could not tell them what they really needed to know. If you look at these issues from the standpoint of service obligation, I suspect that 90% or more of neurosurgeons, at least 90–95% of the time, are satisfied at present with what the pathologist tells them. If I am wrong, I would like to stand corrected.

From the intellectual point of view, I think this is a very interesting, and very necessary discussion. There are many valid points on both sides of the question that impinge on clinical decisions. As Darell Bigner mentioned, we have got to go where things direct us to go. If we do not, some other group is going to do it for us. That is the basic concern.

CHAPTER III

Classification of Pediatric Tumors - Alternative Strategies

FLOYD GILLES, *Department of Neuropathology, Childrens Hospital of Los Angeles, Los Angeles, California*

I would like you to turn your attention from the pathogenesis that we have been hearing about so far, to prognosis, and specifically, toward methods of gaining quantitative information about the prognosis associated with specific individual histologic features, specific clusters of histologic features and as a separate entity, specific clusters of children with differing distributions of histologic features.

We have known as pathologists, since 1912, that differing histologic features are associated with different outcomes. We have also known that some histologic features play no role in the outcome of the child with a particular tumor.

One such specific histologic feature is a microcyst filled with protein material. It is associated with a very good outcome on a simple univariant analysis comparing all children with posterior fossa tumor and that feature to all of those children with tumors in a similar location without that feature. Conversely, other histologic features, such as the perivascular pseudorosette, are associated with an adverse outcome, again using the simple univariant analysis.

Some histologic features within a group of tumors, for instance the Homer Wright rosette, in medulloblastoma - PNET, have a bearing upon outcome. There are differences between these survival curves, and simple univariant analyses, for those medulloblastomas in the posterior fossa containing the Homer Wright rosette, compared with those medulloblastomas that do not contain the Homer Wright rosettes.

Other influences confound the search for quantitative prognostic information associated with individual histologic features or clusters of histologic features. There is an apparent influence of age upon outcome in posterior fossa tumors in children.

There are three data bases to which I have had access. Two are small data bases collected at the Children's Hospital in Boston with the help of Drs. HEDLEY-WHYTE and LEVITON (GILLES et al. 1983). The third is a very large data base (Childhood Brain Tumor Consortium) collected with the help of some of the members of the group attending this workshop, and 28 other individuals (WINSTON et al. 1977).

Our goal is to improve the accuracy of prognosis and to obtain specific quantitative information about prognosis associated with specific histologic features and clusters of histologic features. We hope to gain this information, not by a priori application of classifications, but by the individual histologic features. Specific operational definitions were developed for each histologic feature.

We want to improve prognostic information by increasing the homogeneity of histologic classes, and by use of very specific statistical techniques. We want to consider simultaneously the effects of histologic classes, tumor location, amount of surgical removal, and a large number of other clinical and demographic variables upon prognosis.

We have examined the outcome of clustering algorithms for histologic features, and as a separate exercise, the outcome of clustering algorithms applied to children with different distributions of groups of clusters of histologic features. The prognosis is measured by one of the proportional hazards models, the first of which was popularized in 1972 by Cox, but many multivariant techniques of a similar nature are now available. These models estimate the hazard of death, given the large number of potentially prognostically important variables.

My simple definition of a cluster is an association of two or more histologic features. This is simply a formalization of the process that each of us does individually on a daily basis in our practice of neuropathology. We look for features that are associated. We expect certain combinations of histologic features to appear, and we expect those histologic features to appear in isolation. We do not expect, for instance, patches of medulloblastoma within a pilocytic astrocytoma.

How are these clusters measured? This is a simple measure of an association in a cluster. A cluster refers to a population of children, not to an individual child. One can in a given population determine an expected proportion of cases containing two histologic features, regardless of what they are, by simply multiplying the proportion of each histologic feature in that population.

An example of a cluster was one encountered in 198 children with densely cellular tumors which were not presorted as to whether or not they fit traditional names. Two features which were clustered happened to be the features of necrosis, and macrophages which is certainly not unexpected.

In the description or development of this particular data base, including the 198 children, we used 44 individually, operationally defined histologic features, and 103 clinical variables. Four clusters appeared. One of these with four features is interesting. It contains patches of astrocytoma, patches of low cell density, medium cell density and small patches of high cell density. We were interested in knowing whether this cluster was important in determining outcome for the children containing it.

It is essential to emphasize a separate point, that individual histologic features are not equally important in estimating prognosis. What do we mean by that? When one takes individual histologic features and use a proportional hazards model, one finds that some individual histologic features are associated with a decreased median survival, in conjunction with an estimation by the neurosurgeon of the amount of tumor removed. One finds that other histologic features are associated with an increased median survival, again in conjunction with localization information and the decade during which the operation was performed.

A proportional hazards model is a multivariant analysis that identifies and ranks variables in accordance to their importance in estimating prognosis. Using this densely cellular cerebellar tumor population of 198 children, we compared those children whose tumors contain all of the features of the astrocytoma cluster to those without any of the features of the astrocytoma cluster. The medium survival of those without any of these particular features was nine and a half months. The median survival of those with all of the features was 63 months. The combination of the strategies of clustering and proportional hazards models seemed to give us quantitative prognostic information.

In using the large population from the childhood brain tumor consortium data base, we entered into the proportional hazards model 50 individual histologic features, and a small number of clinical features, such as decade of surgery, amount of tumor removed, what the surgeon's observations were as to the margin of the tumor, presence or absence of papilledema, and the age of the patient. Of the 50 histologic features, only a few were associated with an increased risk of death. Similarly, a relative few were associated with a decreased risk of death.

We applied the clustering algorithms to the material. At the present time, we are working with 13 clusters of individual histologic features. These clusters are defined on the basis of the association among the histologic features within the cluster. We have the quantitative information of the distances apart or the amount of association between clusters. Only a few of these clusters are actually associated with a change in the risk of death. One cluster is associated with a 9-fold decrease in the risk of death. Another cluster was associated with a 2.9-fold decrease in the risk of death. A third and a fourth cluster were associated with an increase in the risk of death. A fifth cluster was associated with a decrease in the risk of death. The proportional hazards model now allows us to rank the individual clusters on the basis of the magnitude of the contribution of each to the proportional hazards model.

The final step in clustering is to provide information about groups of children based on the similarities among each of the clusters. This is conceptually a somewhat more difficult step. It asks the question for each child in the population, "What is the distribution of clusters within the tumor?" Each child has a different value for each cluster. Some clusters of histologic features may be present, while some may be absent, as you would expect in any given child. However, all children have values for each of the 13 clusters of histologic features.

Let me translate that into the reality. We have all encountered pilocytic astrocytomas with a patch of high cell density. What do you do about a patch of high cell density within such a tumor when you provide for the clinician a statement regarding prognosis? Some neuropathologists would rank that tumor as a bad one, and some, I suspect, would say that the presence of a pilocytic component within the tumor provided a somewhat better prognostic outlook, especially if the tumor was limited to the posterior fossa. If the child's tumor contains all features of a pilocytic cluster, it would have a value of *one*. If only half of the histologic features were present, the value would be *one-half* and, if none, *zero*.

When one has this information about each of the children in a large population, and localization information is available, one can assign to each child a value for each of the 13 individual clusters. The clustering algorithm can then take a group or cluster of children on the basis of the individual features.

Each histologic feature or clinical feature or cluster of features contributes a value to each child's tumor. That means that the value may be *zero* if all features are absent and may be *one* if all features are present. If only some of the histologic features are present, the value may be intermediate. The clustering algorithm provided some 35 different groups of children.

What we want to do and have not been able to do yet, is to construct proportional hazards models on each of these individual clusters of children.

What I have introduced you to in this brief presentation is a series of measures for providing quantitative prognostic information to the clinician about prognosis derived from association with each of many histologic features or constellations of histologic features.

TABLE I Distribution of histologic features in 132 glial tumors of the cerebellum and the survival for children with each feature (WINSTON et al. 1977)

		Life table			
Histologic feature [a]	Percent of cases with feature	Fraction alive one-half mo operation	Fraction alive 1 yr from operation	Fraction alive 5 yr from operation	Fraction alive 10 yr from operation
Focus of oligodendroglia	12.1	1.00	1.00	1.00	1.00
Microcyst	56.8	0.99	0.96	0.92	0.90
Rosenthal fiber	28.0	0.95	0.92	0.89	0.86
Leptomeningeal deposit	32.6	0.95	0.88	0.84	0.81
Low cell density	44.7	0.92	0.88	0.81	0.78
Pleomorphism	28.0	0.95	0.81	0.76	0.76
Hemosiderin	6.1	1.00	0.75	0.75	0.75
Endothelial proliferation	52.3	0.96	0.81	0.78	0.75
Astroglial cells	100.0	0.93	0.81	0.72	0.69
Medium cell density	47.7	0.97	0.78	0.71	0.67
Hypervascularity	26.5	0.89	0.77	0.65	0.65
Calcification	22.7	0.90	0.70	0.57	0.50
Parenchymal desmoplasia	7.6	1.00	0.80	0.60	0.48
Necrosis	14.4	0.90	0.63	0.47	0.47
Perivascular desmoplasia	12.1	0.94	0.62	0.56	0.42
Mitosis	6.8	0.78	0.56	0.33	0.33
High cell density	7.6	0.80	0.60	0.30	0.30
Perivascular pseudorosette	14.4	0.84	0.42	0.26	0.26
Ependymal rosette	3.0	0.75	0.25	0.25	0.25

[a] Features are arranged in order of increasing risk of death at 10 yr from operation. Two non-tumor-related deaths were treated as lost to follow-up. One death attributed to tumor occurred greater than 10 yr after operation.

Discussion of Chapter III

DR. KLEIHUES: In the beginning, you showed a tumor with typical perivascular rosettes and you showed a graph which indicated that children with tumors containing those rosettes were doing much better than the others.

DR. GILLES: No. I think I went over that too rapidly. The first group of children and the first two pictures that I showed were from a group of 132 children with hypocellular tumors which were limited to the cerebellum. In that group of 132 children, those with a patch of microcysts had an outcome that was very good. Those children with a patch of perivascular pseudorosettes had an outcome that was very bad, just the opposite.

DR. KLEIHUES: Was this tumor then an ependymoma in your view, or an astrocytoma?

DR. GILLES: The traditional classification for the latter tumor would be *ependymoma.*

DR. KLEIHUES: So what you are saying conceptually is that the astrocytoma does better than the ependymoma. Is that the sense of what you are saying?

DR. GILLES: No, that is not the sense. There is no way that you can take a tumor characterized by this one histologic feature, perivascular pseudorosette, and call it a low-grade tumor, grade I or grade II prognostically, unless you measure the outcome. We now know that 80% of the children are going to die.

DR. KEPES: Did I understand you correctly when you mentioned that proximity to the surface is a negative prognostic sign?

DR. GILLES: Yes.

DR. KEPES: In some of your earlier publications, you mentioned that the involvement of the leptomeninges by tumor cells is a positive sign. Now, how can you have involvement of the leptomenges, unless you are close to the surface?

DR. GILLES: I apologize for some confusion in my earlier reports. The first study was a study of 132 children without a patch of dense cellularity (WINSTON et al. 1977). In that group of children, leptomeningeal involvement by the tumor was associated with a very good outcome (see Table I).

The second study was a separate one of a different population of 198 children, with a densely cellular tumor of the cerebellum. In that study the statement about the proximity to the surface of cerebellum, which was a surgical observation, came to the fore.

DR. SZYMAS: May I ask you why you did not use the classical histologic diagnosis in your study? One of the parameters which you used was the proliferation of blood vessels and this proliferation is quite different in pilocytic astrocytoma. For example, in medulloblastoma the proliferation of blood vessels correlates with malignancy, but it does not in pilocytic astrocytoma.

DR. GILLES: That turns out to be quite accurate. When we evaluate separately the presence or absence of endothelial cushions, endothelial proliferation or glomeruloid formation of small blood vessels, we find that endothelial proliferation may also be present. However, even if it is present, there is usually a very good outcome.

You are quite right, and this is exactly the point I am trying to make, that clustering children on the basis of these variables, the variables make different contributions in different groups of children. What you are making is a very important and fundamental point. In a child with a cluster that might resemble a glioblastoma for instance, the cluster of endothelial changes is associated with a very bad outcome.

This is the basic point I am trying to get across. I want people to think a little bit more about clusters of children when predicting outcome, rather than just thinking in a static way that a cluster of histologic features on a slide may sometimes be associated with a good outcome or a bad outcome.

DR. EISENBERG, Galveston: This is interesting but it seems to me that the validity of the whole business is dependent upon the outcome data. A confounder would be that the decisions for therapy are based on classical nomenclature. How do you get around that? In other words, you are grouping on the basis of clustering, and yet the decisions for therapy are on the basis of naming. Does that present a confounder?

DR. GILLES: Well, at this stage, it would present a confounder. Now, let me tell you about the validity of the outcome. First of all, we have measured the observational variation on each of these features, and in the major study there are 144 histologic features. How accurate are we at seeing the histologic features? We know on which histologic features we have low observational variation. In fact, Dr. Rorke and Dr. Davis did not just look at 3,300 brain tumors. They were subjected to a Machiavellian torture of having to look at some of those tumors a second time. They did not know which tumors they had diagnosed previously because we were assessing their observational variation with respect to putting tumors into the WHO classification. They looked at well over 4,000 individual cases.

In that population, we had a large number of children. As for posterior fossa tumors, we had 1,243 children. In order to ask the question whether or not these clusters have any value or these prognostic statements have any value, we randomly divided those children into two boxes. The statements I have given you today are merely speculative and are hypothetical until we test the validity of these clusters and the prognostic statements.

How are we going to test them? I have given you the clusters, and prognostic statements from the first box. The second box remains closed until we are satisfied that those histologic feature clusters, child clusters, and the prognostic statements about them are valid as well. Then each of the statements will be tested by undertaking exactly the same exercise on the children who are at present in the other closed box. In that manner we hope to determine whether or not the clustering works. We will also learn whether the prognostic statements work in two different populations in which the data were collected in exactly the same fashion.

DR. ZÜLCH: Is not the survival time, and that means biologic behavior of these two tumors which you reported on, defined by the location? Pilocytic astrocytomas are mainly, or almost without exception, located in the midcerebellum, which means the vermis. The ependymoma is very much fixed to the posterior third of the fourth ventricle with all of the very sensitive nuclei of the cranial nerves.

But common practice has shown, and you showed these microcysts which are the beginning of the formation of large cysts, that if you open the large cysts only and do not do anything to the tumor, this may be sufficient for 15 years survival when the tumor begins to be solid. So, the location of the tumor must always be taken into account if you speak about the survival or the biologic behavior of a tumor of the posterior fossa. That is very important.

DR. GILLES: I agree with you that it is very important, and you have made implicit in your statement a criticism of the WHO classification which, unfortunately, ranks tumors grade I, II, III, or IV regardless of location. We can correct that. I realize of course that the WHO classification was based on a democratic decision, and I recognize the players who were involved in it at that time. But the statements I made were based on tumors thought to be limited to the cerebellum. The ependymoma was limited to the cerebellum or the roof of the fourth ventricle.

Now in point of fact, 30.2 percent of tumors in the posterior fossa, all pilocytic astrocytomas, simultaneously involved the cerebellum and brainstem, and they are included in this population.

DR. ZÜLCH: Well, it's very difficult, and if I would report about our previous discussions extensively, you would understand our dilemma regarding the definition of grading. When you have to define differentiated or undifferentiated tumor cells, then you are in the middle of great difficulties.

I believe it is impossible to say whether a cell is differentiated or undifferentiated or embryonal. What we attempted to do was to provide the neurosurgeon with a certain rule of thumb. We knew about the probable biologic behavior of the tumor. We have to distinguish the morphologic definition of biologic behaviors of malignancy from the clinical. The clinical one is dependent on site, particularly site in reference to vital structures of the brain, such as midbrain or medulla oblongata, which have a close proximity to the cerebrospinal fluid pathways. For example, a pilocytic astrocytoma in the aqueduct is a fatal tumor.

DR. GILLES: Yes. Absolutely.

DR. ZÜLCH: If you do not make a distinction and do not introduce a clinical evaluation, then as a clinician, you get insufficient information. This creates difficulty for us, as morphologists. We can learn something from the morphology but the survival time is defined by the clinician. Because he takes a long series of 200 members of a population and sees what he has done and what the morphologist has said, and then he makes his definition of survival time, which is quite different from our definition based on the morphologic appearance of the tumor.

DR. GILLES: You are absolutely right. For each of these studies, we are doing our best to control for location. For each of these studies, the first with 16 histologic features, the second with 44 histologic features, and the third with 144 histologic features, we have developed an operational definition for each of the histologic features. We have not used such subjective terms as anaplasia, malignancy, differentiation, dedifferentiation; we have used simply descriptive terms.

If you look, for instance, in the neuro-oncology textbooks from the last 25 years, you will see that the term *palisades* is used in some seven different ways indicating that the term has been applied to seven different histologic patterns. We wanted terms that were teachable and could be communicated to our young people. Yet, we wanted these terms to be as specific and understandable as possible. We have tried to follow your dicta and your leadership on these particular points.

DR. KEPES: As Professor Zülch mentioned, I think we all realize how extremely important the localization of the tumor (astrocytoma and other neoplasms) really is for prognosis. The diagram shown below (Fig. 1) was included in a lecture of mine to the Canadian Association of Neuropathologists in 1986, but I think it should be included in this book on brain tumor classification.

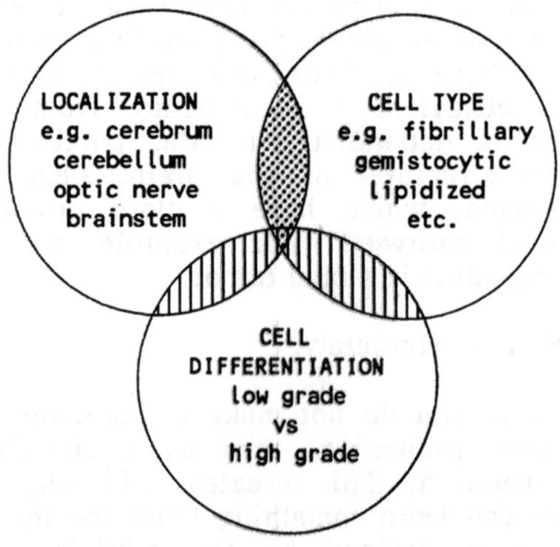

Figure 1 Factors determining prognosis of astrocytomas.

There is one other parameter which is not included here, and that is reponse to treatment. I think that when we make a prognosis for survival, we must consider the response to treatment. If I might use an example from general pathology, a seminoma of the testes is a very primitive tumor, while an embryonal carcinoma is a better differentiated tumor. Yet the latter has a much worse prognosis. A seminoma responds to treatment with a 96 percentile survival, whereas the embryonal carcinoma, in all its glory of better differentiation, has a very poor clinical outcome. So I think we have to consider that.

DR. JELLINGER: I do not understand the biologic rationale for the descriptive terms. It seems to me that you compare, let us say, a human baby with a fish or with a frog, depending on whether he has brown eyes, a size of 20 centimeters, and a weight, let us say, of 0.5 kilograms. It is extremely difficult to follow the terms. You only are entitled to compare the same groups of tumors using the same descriptive criteria, and not combining anything, in order to avoid incorrect statistical data.

DR. GILLES: I am sorry that you feel that way, Dr. Jellinger, but those of us who deal with pediatric brain tumors find the definitions available in the current literature quite fuzzy, particularly in the border zones between the classes of individual tumors. I find that the boundaries between grades I, II, III and IV are very fuzzy.

We are using all kinds of measures, just like numerical taxonomy, and all of its influence on classification of all kinds of scientific objects. We use multiple measures of multiple variables in studies of these children with brain tumors. We do not judge that a tumor which has been derived from some hypothetical embryonic cell in a hypothetical fashion, has a hypothetical outcome. We do not make that prejudgment. At the present time we use the simplest definitions we can.

DR. ARMSTRONG: We have had some very clear definitions of how names have been used in description and diagnosis of pediatric brain tumors. Dr. Rorke has elucidated her classification of primitive neuroectodermal tumors. She has defined it for use with tumors within the central nervous system. She has explained how her definition helps to solve the problems of classification of many childhood brain tumors which do not fit into the present World Health Organization classification. She has told us how it can be used to incorporate new techniques, and in this way define the behavior of the tumors which she defines as undifferentiated primitive neuroectodermal tumors.

Dr. Rubinstein has clarified his use of seven embryonic types of tumors by defining where they are, how they look and how they behave. Moreover, he has shown how he has been able to identify a new tumor using his concept of embryonic differentiation and classification of pediatric tumors.

There has been a plea for the use of new diagnostic markers, and it has been expressed that we are going to have to do this in a prognostic way and not use old material. I think it is important for us to define methods for properly preserving tissue so that we can apply these new techniques.

We have heard a carefully described alternate method of classifying tumors, based on cell apearance and morphology, disregarding names, and incorporating various histologic features, with clinical behavior. We shall look forward to Dr. Gilles' publication of this new approach, which is an alternative to using names for which we have no concensus.

Please keep in mind the following questions: (a) Is the World Health Organization classification adequate for your purposes in diagnosing childhood tumors? (b) Is the proposed alternate classification a useful one, or could we define better what constitutes a useful classification?

CHAPTER IV

Primitive Neuroectodermal Tumors: Views on a Working Classification

LAURENCE E. BECKER, *The Hospital for Sick Children and University of Toronto, Toronto, Ontario, Canada.*

Introduction

The classification of brain tumors began many years ago with BAILEY and CUSHING (1926). Surprisingly few fundamental alterations have been made to their system of nomenclature, although numerous modifications have been suggested. Because of the proliferation of individual classifications, in 1979 the World Health Organization (WHO) sponsored publication of histologic typing of central nervous system tumors produced by a consensus of neuropathologists (ZÜLCH 1979). This system, based on hematoxylin-eosin staining (H&E), was expected to be a working classification. In 1985 a modification of the WHO classification was suggested, reflecting the views of neuropathologists dealing primarily with brain tumors in children (RORKE et al. 1985).

Although differences of opinion about classification of non-neurologic tumors exist (e.g., lymphomas), the most heated recent controversies have erupted over brain tumors. In this presentation we will consider a working classification of primitive neuroectodermal tumors. These tumors represent a significant portion of childhood brain tumors (PARKER et al. 1975), and brain tumors are the second most frequent tumor in children, after leukemia. Of the 1,400 patients in the Brain Tumor Registry at The Hospital for Sick Children in Toronto, about 20% have primitive neuroectodermal brain tumors (85% infratentorial and 15% supratentorial) (BECKER and HALLIDAY 1987).

The term *primitive neuroectodermal tumor* has generated numerous editorials, personal letters, phone calls, rebuttals, and workshops (BECKER and HINTON 1983; RORKE 1983; RUBINSTEIN 1985: DEHNER 1986: TRICHE 1986; DEHNER 1987; RUBINSTEIN 1987b; TRICHE 1987). This excitement is not a result of our great knowledge about this group of tumors, but rather of our ignorance. Because of our lack of knowledge, I believe that we must have a practical approach to nomenclature in this significant group of childhood brain tumors. We need a working classification.

The Purpose of Classification

To be effective, the names used in the classification must convey a consistent message about the histologic appearance of the tumor, within the limitations imposed by our lack of knowledge about basic tumor biology. The hospital neuropathologist names a tumor to communicate to the responsible physician the essence of what he has seen on examining the surgical specimen. This information is then transmitted to a wide range of persons, including the patient, the oncologist, and other neuropathologists. Based on the pathologic diagnosis, a prognosis is given to the patient and family members and a treatment protocol is established.

The diagnostic labels applied to tumors must also have scientific validity. Although molecular biology is rapidly changing our view of many conceptual aspects of neuro-oncology, molecular techniques are limited to a small fraction of laboratories. Thus, terms may be scientifically sound (based on the most recent molecular biology) but difficult to apply in the average hospital laboratory (Donner et al. 1985). In contrast, pathologists may be able to consistently diagnose a type of tumor when molecular biologists can detect no common genetic theme. Therefore, the working classification by necessity lags behind scientific advances. Nevertheless, by adhering to strict rules of observation rather than on suppositions or hypotheses about tumor cell origin, application of diagnostic terminology will achieve greater longevity and wider acceptance. For example, the concept that each tumor type is related to a specific stage of brain maturation is very attractive (Rubinstein 1985b). However, scientific facts about tumor cell origin are scarce. A relationship to embryonic development may exist but the evidence is not overwhelming at this time (Rorke 1983).

Limitations of Diagnosis

The ideal investigation of tumor tissue may be possible at only one or two institutions in the world and for only one or two cases per year. Many practical limitations are present. The amount of tissue available for processing is important: generally, the more tissue the better. The manner of tissue fixation (e.g., frozen sample only, fixed for electron microscopy) determines the types of test that can be undertaken. Time is also a factor: the neurosurgeon needs instant diagnosis of a frozen section, the family is anxious about their child, and the oncologist wants to initiate treatment immediately. Opportunities for unlimited testing and wide consultation often are not practical to achieve an appropriate working diagnosis. However, from a scientific point of view, further investigation and additional consultation can continue for decades. The tests available may be restricted by the antisera in stock in the laboratory and budget constraints limiting the number of tests that can be made and determining whether electron microscopy is performed. The pathologist examining the tissue is another variable, which could be eliminated only if all brain tumors were seen by *the* world expert.

In summary, the neuropathologist usually must compromise his investigation in a variety of ways. Nevertheless, he aims to use the tissue he has to examine and the facilities at hand judiciously to produce an appropriate, helpful diagnosis. The useful classification system tends to be the one that is meaningful, reproducible, quickly achieved, and inexpensively confirmed.

The neuropathologist applies the basic concept that tumors of similar appearance are given similar names. However, tumors of similar histologic appearance may have a surprising heterogeneity of cell types. With increasing information about the antigenic structure of cells, the diversity of these tumors becomes apparent (DONNER et al. 1985).

In real life the pathologist is hampered by his lack of a complete spectrum of oncologic techniques to apply in the diagnostic process. To emphasize the difference between a service and a research laboratory the potential investigation on a surgical specimen has been arranged in a hierarchical order of sophistication (Table I). The majority of service laboratories are first or second order; very few are beyond the fifth order. Therefore, a useful working classification must be applicable to first- and second-order laboratories. My preference is to see a classification based on the fourth order, including electron microscopy and extensive use of immunohistochemistry. This Table emphasizes the degree to which most neuropathologists are limited in their utilization of the scientific background when going through the process of making a diagnosis.

The Controversy over the Term Primitive Neuroectodermal Tumor

Drs. Rorke and Rubinstein, at this meeting, have comprehensively reviewed the evolution of the concept of primitive neuroectodermal tumor. My perception is that there is no difficulty with the concept of a group of tumors that are neuroectodermal and happen to be primitive. These tumors have a similar phenotypic appearance and a tendency to demonstrate neuroectodermal markers.

Table I. Hierarchy of Tumor Investigation

Order	Investigation
First	H & E
Second	Special stains
Third	Immunohistochemistry (antisera generally available)
Fourth	Electron microscopy
Fifth	Immunohistochemistry (antisera restricted to particular laboratories)
Sixth	Karyotyping and DNA ploidy
Seventh	Oncogene identification
Eighth	Oncogene amplification
Ninth	Gene deletion
Tenth	Nucleotide sequence abnormality
Eleventh	Others

The controversy appears to be whether the name given to a tumor should reflect its embryonic origin. Since the evidence is insufficient to support an embryonic relationship, using a label such as *pineoblastoma* that incorporates an embryologic implication suggests more understanding of the tumor biology than actually exists. Furthermore, using different *blastoma* labels for tumors in different locations leads to errors. We have many examples in which it cannot be ascertained whether the tumor arose from the pineal or the cerebellum. For instance, would the tumor in Figure 1 be called *pineoblastoma, neuroblastoma*, or *medulloblastoma*? Or would the pathologist be tempted to invent a term? Therefore, many pathologists regarded the introduction of the diagnostic label *primitive neuroectodermal tumor* by HART and EARLE (1973) as appropriate.

Previously I proposed that the term *primitive neuroepithelial tumor* be used to describe the concept and *primitive neuroectodermal tumor* be the specific name for a tumor (BECKER and HINTON 1986). In practice, over the past few years I have used the label *primitive neuroectodermal tumor* for those that occur in the cerebrum, cerebellum, and pineal gland and have similar histology. This habit has evolved because histologic, immunologic, and ultra-structural examinations generally fail to distinguish primitive neuroectodermal tumor of the cerebrum and pineal from those of the cerebellum. Although the use of different terms for the concept and the diagnostic label is logical, it is difficult in practice. Thus, the term primitive neuroectodermal tumor tends to be used for both.

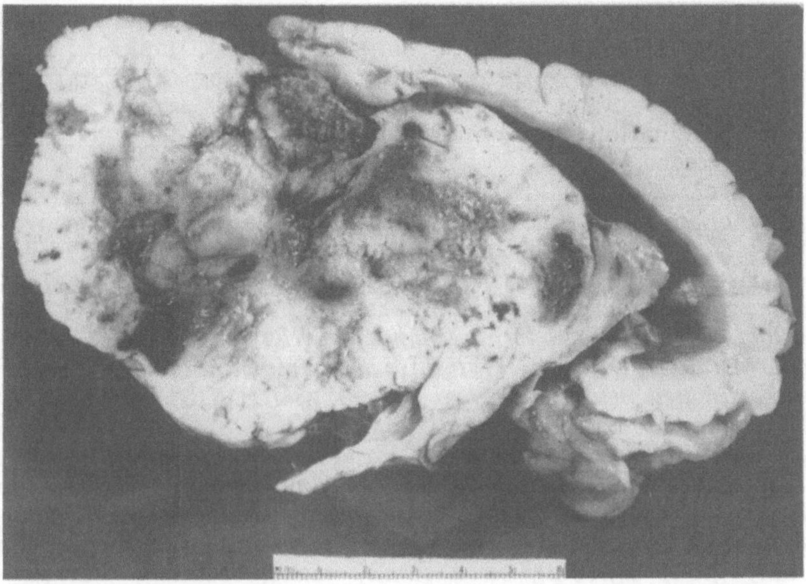

Figure 1. Sagittal section of a large, congenital, primitive neuroectodermal tumor with widespread infiltration.

Agreement on the family name for this group of tumors is important but difficult to achieve. Often the pathologist justifies the use of a unique designation on the basis of minor histologic variations that have no apparent impact on prognosis. For tumors not easily classified histologically, the tendency is to create new names (VANDENBERG et al. 1987b), which is not surprising since tumor names are largely based on histologic appearance.

The types of names given to these tumors reveal the wide range of confusing terminology that has evolved. For example, a tumor with melanin may be designated *melanotic mucin-producing neuroepithelial neoplasm* (SHUANGSHOTI 1980), *melanotic progonoma of the brain* (STOWENS and LIN 1974), *medulloblastoma-like tumor with melanin formation* (BEST 1973), *melanotic medulloblastoma* (DRUT and JONES 1983), *malignant melanotic neuroectodermal tumor* (DEHNER et al. 1979), *melanotic ameloblastoma (progonoma)* (WILLIAMS 1967) or *embryologic tumor of retinal anlage* (CLARKE and PARSONS 1951). None of these terms is better than *primitive neuroectodermal tumor* (family name), *with melanin* (given name).

Some neuropathologists never use the term *primitive neuroectodermal tumor* (RUBINSTEIN 1985b), while others use it to describe any medulloblastoma-like tumor occurring at any site within or without the central nervous system (DEHNER

1986). Between this spectrum are those who restrict the designation of *primitive neuroectodermal tumor* to specific sites (RORKE 1983; BECKER and HINTON 1986). This use of the term places tumors of similar histology, immunochemistry, and differentiating potential in a coherent group, making them accessible for determining biologic behavior. In contrast, if these tumors remain as rare, isolated cases, often with a unique designation, our understanding is reduced to anecdotal experience.

However, we should be open-minded about the label we actually choose for this group of tumors. Perhaps most neuropathologists would feel more comfortable with a different acronym, which could easily be substituted for *PNET* (*primitive neuroectodermal tumor*). For example, we could have *BNET (blastomatous neuroectodermal tumor), ENET (embryonal neuroectodermal tumor), PNEPT (primitive neuroepithelial tumor), USNET (unsophisticated neuroectodermal tumor)*, or *IUDNET (intrinsically undifferentiated neuroectodermal tumor)*; or, as a compromise, we might use *RRNET (Rubinstein and Rorke's neuroectodermal tumor).*

Controversy over names of tumors about which we know so little can become ludicrous. However, it would be an accomplishment if we could come away from this workshop as a friendly group agreeing on a family name, realizing that given names for members of this family will continue to vary, perhaps depending on the personality of the pathologist, the amount of tissue available for examination, and other factors.

A commonly accepted family name is practical for designing treatment protocols and ensuring appropriate management of patients. The treatment of many brain tumors is not particularly effective and treatment protocols are limited. Therefore, most oncologists tend to devise regimens on the basis of tumor families and sites rather than on specific given names. Thus, when tissue is limited, the diagnosis may be primitive neuroectodermal tumor. However, if a great deal of tissue is present, other cell types (e.g., ependymal cells, neural tube cells) may be identified that would allow a given name to be added. The given name then would convey additional information about the other cells. As more sophisticated treatments are developed, the presence of other cell types will influence choices.

The term *primitive neuroectodermal tumor* has the advantage that *primitive* does not necessarily imply embryonic origin. The concept of differentiation is also minimized by using primitive neuroectodermal tumor, which allows us to use diagnostic labels objectively, without preconceived assumptions, and without indicating the stem cell of origin. Scientific verification of the cell of origin is not available at this time (RORKE 1983; DEHNER 1986; TRICHE 1986). We have no evidence to say whether it is an immature stem cell or a mature cell.

In the Brain Tumor Registry in Toronto we have two categories of primitive neuroectodermal tumors. One category is primitive neuroectodermal tumor with no evidence of other cell types. Site designation is emphasized and the old terminology is given in brackets. The second category is primitive neuroectodermal tumor with the cell type or types identified in the tumor. Again, the site designation is included and the old or other terminology is added in brackets. As specific cell markers become available, cell types will be documented more objectively but this framework will remain useful.

This classification does not arbitrarily lump tumors together. There are at least three discriminators: the presence of primitive cells, identification of other cell types, and designation of tumor location. Some pathologists consider the use of such nomenclature a backward step; however, I think we are at the stage where one step backward is necessary to move two steps forward. For example, there may be a medulloblast but no one has yet identified such a cell (RORKE 1983). We should resume use of the term *medulloblastoma* only if the cell is found.

Examples of how I classify specific tumors found in children follow. This presentation will involve the central primitive neuroectodermal tumors and largely exclude the peripheral primitive neuroectodermal tumors (DEHNER 1986).

Primitive Neuroectodermal Tumor

Primitive neuroectodermal tumors occur most commonly in the vermis of the cerebellum but can also be present in the cerebrum, including the pineal region (PARKER et al. 1975; RORKE et al. 1985; DONNER et al.1985; TRICHE 1987). They characteristically spread through the subarachnoid space and can metastasize outside the nervous system. Response to conventional radiotherapy depends somewhat on tumor location.

Primitive neuroectodermal tumors without the identification of other cell types consist of primitive neuroectodermal cells that contain small, round-to-oval nuclei with a narrow rim of cytoplasm. Sometimes, there are larger cells with round-to-oval nuclei and more obvious cytoplasm, which may or may not be tapered to one side. Nuclei have prominent chromatin and usually one or two nucleoli. Mitotic figures are readily identified. The intercellular background may or may not have a pink neuropil appearance. Areas of necrosis are often present and vascular proliferation may be noted. These tumors have a variable fibrovascular stroma. Large reticulin-free islands of tumor cells may be surrounded by prominent bands of connective tissue. Homer-Wright rosettes may be identified. These tumors show no positivity with antisera to glial fibrillary acidic protein (GFAP) or neurofilament protein.

Ultrastructurally, the cells are in close apposition, often with interdigitation of cytoplasmic processes but without evidence of specialized intercellular junctions (BOESEL et al. 1978; KOSNIK et al. 1978; MARKESBERY and CHALLA 1979). The nuclei are prominent and contain dense chromatin along their margins. The cytoplasm has sparse organelles but mitochondria and many free ribosomes are usually identified.

Primitive Neuroectodermal Tumor, with Astrocytes

Primitive neuroectodermal tumors with astrocytes may be associated with many histologic patterns. Astroglial cells may be readily apparent or evident only with the application of antisera to GFAP (CAMINS et al. 1980; MANNOJI et al. 1981; PALMER et al. 1981). When the astroglial cells are apparent they tend to be packed into lobular areas in which the intracellular space has a neuropil appearance. Often these lobular areas are sharply defined by reticulin.

Cells are identified as astrocytes by their positivity with antisera to GFAP. They also have the ultrastructural appearance of astrocytes (CAMINS et al. 1980). When the astrocytes are not so obviously delineated, the GFAP may identify groups of cells that have a GFAP-positive capping toward one side of the cell. Astrocytic cells that are part of the tumor must be distinguished from those that are merely reactive to the presence of the tumor (BECKER and HINTON 1986). Reactive astrocytes are often prominent around blood vessels and tend to have numerous processes evident on both H&E and immunoperoxidase staining with GFAP.

Primitive Neuroectodermal Tumor, with Neuronal Cells

Primitive neuroectodermal tumors with neuronal cells may have immature or well-defined neurons (AHDEVAARA et al. 1977: RHODES et al. 1978; CAMINS et al. 1980; DASTUR 1982; BERGER et al. 1983). It is important to distinguish pre-existing neurons from those that are a component of the tumor. Well-defined neurons that are part of the tumor often appear in clusters, frequently with their cytoplasmic borders touching. The cells have vesicular nuclei and prominent nucleoli. The well-defined neurons are positive with antisera for neurofilament protein when the immuno-peroxidase method is used. Ultrastructurally, they have synapse formation, clusters of clear microvesicles, and dense core vesicles. Some investigators believe that the blunt cytoplasmic processes with abundant actin filaments and microtubules resemble growth cones of developing nerve cells (BIGGS and POWERS 1984). More frequently, histologic features are very subtle or absent, but early neuronal differentiation is indicated by positive reaction with antisera to neuron-specific enolase (BECKER and HINTON 1986) or synaptophysin.

Primitive Neuroectodermal Tumor, with Oligodendroglia

Primitive neuroectodermal tumors with oligodendroglia are not well documented largely because of the absence of well-established cell markers (BECKER and HINTON 1986). However, rarely cells resembling oligodendroglia with the typical perinuclear halo appearance are identified.

Primitive Neuroectodermal Tumor, with Ependymal Cells

Primitive neuroectodermal tumors with ependymal cells are the equivalent of *ependymoblastoma* (DOHRMANN et al. 1976; MANNOJI and BECKER 1988). They are composed of fields of immature cells in which ependymal rosettes or canals can be identified (DOHRMANN et al. 1976). These features may be extremely rare or quite abundant. The immunoperoxidase reaction with GFAP may or may not be positive. Cytokeratin positivity has been noted in some cases (MANNOJI and BECKER 1988). Electron microscopy shows complex junctional devices, microvilli, cilia, and basal bodies. Mitotic figures are common and may be found within the rosettes. The rosettes also may have multiple cell layers rather than the single layer characteristic of the anaplastic ependymoma.

Primitive Neuroectodermal Tumor, with Melanin Pigment

Primitive neuroectodermal tumors containing melanin pigment are rare and are associated with foci within the tumor that show prominent pigmentation. They may be associated with papillary and tubular formations. The melanin-containing cells tend to be low columnar or cuboidal but do not always form tubular structures and may be scattered in regions of the tumor composed of primitive cells (DRUT and JONES 1983).

Primitive Neuroectodermal Tumor, with Muscle Elements

Primitive neuroectodermal tumors with muscle elements, either smooth or striated, are rare (SMITH and DAVIDSON 1984). The striation within the muscle fibers can be suggested by the phosphotungstic acid hematoxylin stain and confirmed by immunohistochemistry and electron microscopy. The nature of these tumors is controversial. Various investigators have suggested that they are teratoid tumors, mixed epithelial mesenchymal neoplasms, or primitive neuroectodermal tumors arising from the mesoectoderm (AUER and BECKER 1983; SMITH and DAVIDSON 1984).

Primitive Neuroectodermal Tumor, with Fleurettes and/or Flexner-Wintersteiner Rosettes

Most primitive neuroectodermal tumors of the pineal region in children do not show Flexner-Wintersteiner rosettes or fleurettes (SOBEL et al. 1981; BECKER and HINTON 1986). The presence of rosettes or fleurettes is suggestive of photoreceptor differentiation (STEFANKO and MANSCHOT 1979). The rosettes consist of a ring of cuboidal cells surrounding a central lumen of varying size. At the apex of these cells is a limiting membrane, which on electron microscopy consists of a series of terminal bars similar to complex cell junctions. Fleurettes are groups of tumor cells with relatively long cytoplasmic processes extending beyond the fenestrated membrane. Some of these processes fan out suggesting the appearance of a bouquet of flowers. Ultrastructurally, cilia with laminated membranes at the tips may be identified. The presence of retinal S-antigen immunoreactivity (BONNIN and PARENTES 1988) in primitive neuroectodermal tumors of the cerebellum shows that primitive neuroectodermal tumors at sites other than the pineal region may show retinal antigens.

Primitive Neuroectodermal Tumor, with Neural Tube-like Structures

Primitive neuroectodermal tumors with neural tube-like structures have been described as medulloepithelioma located in the cerebrum, cerebellum, brainstem, eye, and peripheral nerve (POLLAK and FRIEDE 1977). The histology is characterized by structures that appear to recapitulate the embryonic neural tube although antigens common to the fetal neural tube and the tube-like structures have not been identified. The prominent epithelium, in the form of tubules, ribbons, or papillary formations, is pseudostratified with a high mitotic rate. Mitoses are more numerous on the luminal side of the tubule. Ultrastructurally, the tumor cells have little cytoplasm and few organelles (POLLAK and FRIEDE 1977). They rest on a distinct basement membrane and have no cilia or microvilli. The tubular cells may merge into sheets of primitive cells without clear differentiation. Almost half of the reported cases with these neural tube-like structures show other cell types, including neurons, astrocytes, oligodendroglia, and ependymal cells (DECK 1969; KARCH and URICH 1972; AUER and BECKER 1983).

Conclusion

Since labelling or changing the label of a tumor does not increase our understanding of its biology, I hope that in future we can concentrate on the fundamental knowledge of histogenetic relationships in this group of tumors rather than the nomenclature. As TRICHE (1986) has suggested, we should be seeking answers to questions such as: Is rearrangement, amplification, deregulation, or increased expression of one or more oncogenes associated with one or more of the central PNETs? Is there a reproducible chromosomal abnormality associated with this group of tumors? What is the *in vivo* and *in vitro* intrinsic differentiation capability of each of these tumors?

I look forward to the elucidation of these answers and the development of the ideal classification based on specific gene expression. In the meantime, although it is not ideal, the designation primitive neuroectodermal tumor provides a working basis for this group of tumors. As DEHNER (1986) has said, "it is a useful scheme to organize our understanding and reveal our shortcomings."

Acknowledgement: This manuscript was prepared with the assistance of the Medical Publications Department, The Hospital for Sick Children.

CHAPTER V

Phenotypic Analysis of Medulloblastoma with Monoclonal Antibodies

DARELL BIGNER, *Preuss Laboratory for Brain Tumor Research, Department of Pathology, Duke University Medical Center, Durham, North Carolina*

The classification of brain tumors discussed in this symposium and the existing WHO classification has been based largely upon a combination of light microscopic findings, anatomic regions in brain in which tumors arise, morphologic similarities of neoplastic cells to developing nervous system cells, combined with clinicopathologic correlations governed by prognostic implications. In contrast to such classification the most meaningful disease classifications are those based upon etiology. Since the etiology and molecular mechanisms of transformation and altered growth control for nervous system tumors remain unknown, it is not likely that it will be possible in the near future to classify brain tumors by either of those criteria. In many other tumor systems from extraneural body sites, particularly with the leukemias and lymphomas, phenotypic classification aided by monoclonal antibody analysis has made major contributions to classification schemata.

The work described in this chapter is based upon an original study conducted by Xuanmin He in D. Bigner's laboratory (He et al. 1989). In the remainder of this chapter those findings will be summarized and placed in perspective. The principles outlined earlier in the Discussion of Chapters I and II for use of monoclonal antibodies were the guiding ones in these studies. That is, all of the work took place with well characterized monoclonal or monospecific polyclonal antibodies which had been shown either to be monospecific and epitope defined, or had been defined operationally by their panel of reactivity on a large number of tumors as well as on normal and fetal human tissues by immunohistochemistry on viable or frozen material. The studies were carried out exclusively in this initial study on four human medulloblastoma lines and xenografts in nude mice and nude rats derived from those cell lines. The rationale for use of cell lines and xenografts consisted of several factors. First, quantitative analyses not possible in tissue, which include percent of cells of a population reacting as in membrane immunofluorescence, and quantitative radioimmunoassays of antibody binding, can readily be carried out with viable cells in culture. Second, one of the major problems in studying phenotypic expression in any brain tumor, particularly when the phenotypes are also represented on normal cells, is determining whether one is looking at neoplastic cells or normal cells because of the infiltrative growth patterns of brain tumors, including medulloblastoma, into normal brain. If one is working with subcutaneous xenografts some of the problems of different phenotypic expression between *in vitro* maintained cell lines and tumor can be overcome and one can be sure that there are no infiltrating normal CNS cells present. Third, the confined growth patterns and the human species specificity of most of the extant monoclonal antibodies allows one to work with intracerebral xenografts as well. Of course, it will be necessary, for final proof and interpretation, to repeat these studies with human biopsy material in order to confirm the findings. Nevertheless, this initial study should provide the basis for continuation of these studies in human medulloblastoma biopsies.

The four cell lines studied that were derived from human medulloblastomas have been described in detail previously. They are TE671 (McAllister et al. 1977), DAOY (Jacobsen et al, 1985), D283MED (Friedman et al. 1985), and D341MED (Friedman et al. 1988). All the lines were cytogenetically monitored and their karyotype, during the testing, corresponded to the published karyotype providing unequivocal identification of the lines and confirmation of their individuality. A large group of monoclonal and polyclonal antibodies were used as detailed in Table I which is reproduced from He et al. (1989) with slight modifications. The antibodies were chosen because they represented several operationally defined phenotypic groupings, namely 1) antineuroectodermal derived tissue-associated; 2) antineuronal cell-associated; 3) antiglial cell-associated; 4) antiextracellular matrix; 5) antiHLA A, B; 6) antilymphoid associated; and 7) anticell surface receptor antibodies. In all cases, both positive and negative control cells and tissues were run simultaneously as were isotype matched control immunoglobulins from myeloma controls or irrelevant antibody controls. Each of the antibodies was titrated and used at optimal dilution for the studies. Four types of assays were used: a cell surface radioimmunoassay, an indirect immunofluorescence assay, immunohistochemistry of xenografts primarily relying on rat xenografts to avoid problems of endogenous mouse immunoglobulin, and immunoblot analysis.

The major results for which the exact experimental findings are listed and detailed in He et al. (1989), are exemplified in Figure 1. The basic phenotype of astrocytic

TABLE I Monoclonal Antibody Profile (adapted from HE et al. 1989)

Antibody	Species Source	Antigen or Immunogen	Reported Specificity	Antigen Location	Concentration Used	Reference or Source
Anti-Neuroectodermal-Associated						
UJ13A	Mouse	Pan-neuroectodermal antigen dermal origin	Normal and tumor tissue of neuroecto-	Cell surface	1.1 μg/ml purified MA	ALLAN et al. 1983
UJ127.11	Mouse	Glycoprotein, 220-240KD	Neuroectoderm, tumors of neuroblastic lineage	Cell surface	12.5 μg/ml purified MA	KEMSHEAD et al. 1983
UJ181.4	Mouse	Neuroblastic cell membrane antigen	Fetal brain, tumor of neuroblastic lineage	Cell surface	5 μg/ml purified MA	COAKHAM et al. 1984
Mel-14	Mouse	Chondroitin-proteoglycan sulfate	Melanoma, glioma, neuroblastoma	Cell surface	1 μg/ml purified MA	CARRELL et al. 1980
15E2E2	Mouse	S-100 protein	Cells and tumors of neuroectodermal, mesodermal and epithelial origin	Intracellular	25 μg/ml purified MA	LOEFFEL et al. 1985
A2B5	Mouse	Gangliosides	Neuronal cells	Cell surface	1/10 ammonium sulfate precipitated protein from culture supernatant	FREDMAN et al. 1984
anti-NSE	Rabbit	NSE	Neuron, reactive astrocytes, astrocytoma, glioblastoma	Intracellular	1/100 antiserum	MARANGOS et al. 1975 PolyScience, Inc.

TABLE I (continued)

Antibody	Species Source	Antigen or Immunogen	Reported Specificity	Antigen Location	Concentration Used	Reference or Source
Anti-Neuronal-Associated						
anti-NFP*	Rat and mouse	Cow, rat & human NFP NFP.150K,NFP.200K	Neuron	Intracellular	1/10 culture supernatant or 1/1000 ascites	CARDEN et al. 1985 LEE et al. 1987 LEE et al. 1986a LEE et al. 1986b SCHMIDT et al. 1987
Anti-Glial-Associated						
anti-GFAP	Mouse	GFAP	Normal, reactive and neoplastic astrocytes	Intracellular	5 μg/ml purified MA	PEGRAM et al. 1985
anti-GS	Rabbit	GS	Glial cells, astrocytoma	Intracellular	1/100, 1/200 anti-serum	PILKINGTON et al. 1982
C12, D12	Mouse	D-54 MG cell	Fetal brain, tumors of neuroectodermal origin except neuroblastoma	Cell surface	1 or 5 μg/ml purified MA	WIKSTRAND et al. 1986
B4, B5, E9 D5, H10 C10+	Mouse	D-54 MG cell	Glioma	Cell surface	1-10 μg/ml purified MA	WIKSTRAND et a., 1986
Anti-Extracellular Matrix						
81C6	Mouse	Tenascin	Glioma, fibroblasts, hepatic and splenic sinusoids	Extracellular	5 μg/ml purified MA	BOURDON et al. 1983
2A6	Mouse	Tenascin	As for 81C6	Extracellular	5 μg/ml purified MA	McCOMB et al. 1984

TABLE I (continued)

Antibody	Species Source	Antigen or Immunogen	Reported Specificity	Antigen Location	Concentration Used	Reference or Source
Anti-HLA						
anti-HLA-A,B	Mouse	HLA-A,B	Human HLA-A,B public specificites	Cell surface	5 μg/ml purified MA	Hybritech, Inc.
anti-B2-microglobulin	Mouse	B2-microglobulin	Human B2-microglobulin	Cell surface	2.5 or 5 μg/ml purified MA	Hybritech, Inc.
L-203	Mouse	Human B cell line 8866	Human Ia molecule, glioma, osteosarcoma	Cell surface	1/250 ascites	LAMPSON and LEVY, 1980
Anti-Lymphoid Marker						
anti-CALLA		Human CALLA	Non T acute lymphoblastic leukemia, B and T cell tumors and peripheral blood granulocytes	Cell surface	2.5 μ/ml purified MA	Becton Dickinson, Inc.
OKT-8	Mouse	31 KD cell surface antigen on T cell	Suppressor/cytoxic T cells	Cell surface	2.5 or 5 μg/ml purified MA	Ortho Diagnostic Systems
390	Mouse	Thy-1 molecule	Thymocytes, brain cells, fibroblasts, glioma, neuroblastoma, rhabdomyosarcoma, leiomyosarcoma, teratoma	Cell surface	1 μg/ml purified MA	SEEGER et al. 1982
anti-Tac	Mouse	Receptor for T cell growth factor	Activated T cells	Cell surface	1/250 ascites	LEONARD et al. 1982

TABLE I (continued)

Antibody	Species Source	Antigen or Immunogen	Reported Specificity	Antigen Location	Concentration Used	Reference or Source
Anti-Receptor						
454A12	Mouse	Transferrin receptor	Proliferating cells	Cell surface	2.5 µg/ml purified MA	FRANKEL et al, 1985
528	Mouse	EGF-receptor	Epithelial cells and tumors, fibroblasts: brain tumors of non-neuronal origin	Cell	2.5 or 10 µg/ml	KAWAMOTO et al, 1983

* The anti-NFP MAs used here were produced by LEE et al. (1986a; 1987) using rat, cow and human NFP rich extracts as immunogens. The specificity of these MAs for human NFP, which have been shown to differ from their specificity for NFP of other species (LEE et al. 1986b), have been determined as described earlier (LEE et al. 1986b; SCHMIDT et al. 1987).

+MA C10 has not been published, but was derived from same fusion as B4, B5, D5, E9 and H10.

MA = Monoclonal antibody; KD = Kilodalton; NSE = neuron-specific enolase; NFP = neurofilament protein; GFAP = glial fibrillary acidic protein; GS = glutamine synthetase; CALLA = common acute lymphoblastic leukemia antigen; EGF = epidermal growth factor.

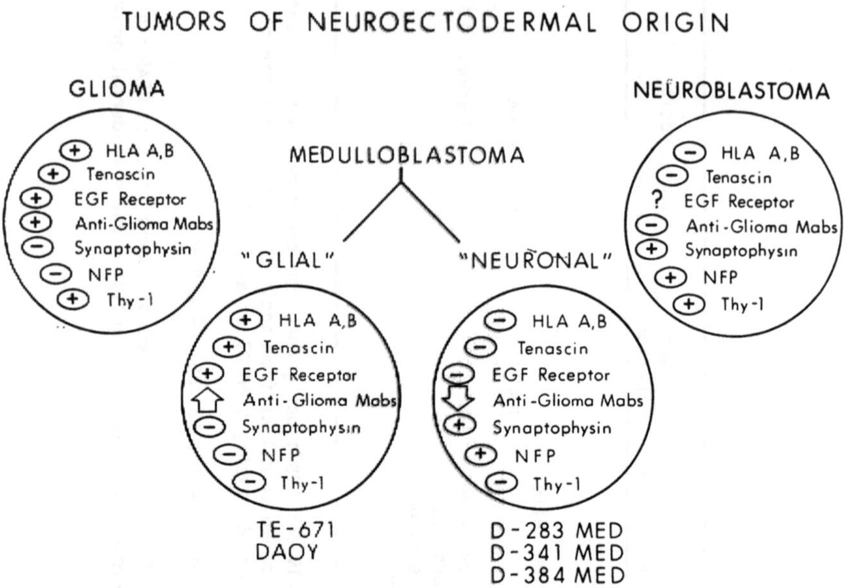

Fig.1 General pattern of reactivity of a broad spectrum of monoclonal antibodies against glioma-, medulloblastoma-and neuroblastoma-derived lines. NE= neuroectodermal-associated antigens; NFP= neurofilament protein; EGF= epidermal growth factor; ND= not done.

gliomas is exemplified in the panel in the circle to the far left designated *gliomas*. Gliomas typically express neuron-specific enolase, are negative for neurofilament protein, are positive for most of the group of antiglioma antibodies including glial fibrillary acidic protein, and they almost uniformly express tenascin, HLA-A, B, and beta 2 microglobulin, the EGF receptor and Thy-1. On the other hand, as exemplified by the test panel shown to the far right for peripheral neuroblastoma, neuroblastomas express neuron-specific enolase, the neuroendocrine marker synaptophysin, neurofilament protein (NFP) and other neurotransmitter enzymes but fail to react significantly with the antiglioma panel, particularly glial fibrillary acidic protein (GFAP), and fail to express tenascin and HLA-A, B and have low to non-detectable levels of the EGF receptor, although Thy-1 is rather uniformly expressed.

To our surprise in these findings, the four medulloblastoma lines fell into two rather distinct phenotypic groups. DAOY and TE671, although not identical to the glioma phenotypic profile, resembled it rather closely. These two cell lines and their xenografts reacted with neuron-specific enolase, failed to react with neurofilament protein, reacted with the majority of the antiglioma antibodies, although negative for anti-GFAP, and expressed significant amounts of tenascin, HLA-A, B, and EGF receptor. They were negative with Thy-1 antibodies. On the other hand, D283MED and D341MED, both of which are suspension lines, contained neuron-specific enolase, reacted at least with some of the neurofilament triplet protein antibodies, showed little reactivity with the antiglioma panel, and were uniformly negative for tenascin, HLA-A, B, the EGF receptor antibodies and Thy-1.

In addition to providing a provisional separation of these four medulloblastoma lines into those with predominantly *glial* and predominantly *neuronal* phenotypes, some points bearing on the discussion with regard to so-called primitive neuroectodermal tumors came forward. All three of the anti-neuroectodermal antibodies produced by John Kemshead and colleagues, namely, UJ13A, UJ181.4, and UJ127:11 reacted with all of the lines with the exception of UJ181.4 failing to react with TE671. Another potentially interesting finding of perhaps practical importance for therapeutic considerations is the reactivity of Mel-14 which reacts with a chondroitin sulfate proteoglycan molecule. Mel-14, which reacts significantly with melanomas and astrocytic gliomas, reacted only with TE671 and DAOY but only minimally with one (D283MED) with the neuronal group of tumors. The results of of the neurofilament protein expression in D283MED and D341MED were confirmed with immunoblot analysis. This unequivocally demonstrated the neuronal phenotype of these two lines.

We view these results as tentative and provisional but as potentially useful in identifying presently available antibodies and suggesting lines of approach for other markers to determine whether this subdivision of medulloblastoma into *glial* and *neuronal* phenotypes is warranted. A much broader battery of neuronal markers, including neurotransmitter enzymes and gene probe analysis for other types of neuronal markers should now be done.

No attempt in this study could be made or was contemplated for classification in a prognostic sense since these were simply cell lines. Several of the initial studies attempting phenotypic classification that have been undertaken with medulloblastoma biopsies have failed to have prognostic significance. We suspect that this failure may result from inaccurate phenotyping, as we have discussed above, and from use of tissue that either had undergone autolysis prior to tissue processing, or that had significant antigenic and epitope loss from fixation with aldehydes or formalin. Our results illustrate clear endpoints obtained in studies with viable or fresh-frozen cells. This type of analysis may very well then provide correlations with ordinary light microscopic criteria with which a new classification scheme, based on molecular criteria, can be considered and tested vigorously for usefulness and validity.

Discussion of Chapters IV - V

DR. JÄNISCH: I would like to make some remarks regarding Dr. Becker's presentation. First of all, I have some doubts whether it is reasonable to have a special classification for brain tumors in children. I think all tumor types which we see in children can also be seen in adults. That includes the primitive neuroectodermal tumors, or as we call them, the *stem cell* tumors. We collected 62 cases from our own material, and 7 of them were in adults. Therefore, I think there is no need to have a special classification for brain tumors in children.

The second thing is, how far should we go in classification in saying what features we see in the tumor. I think we have to make a distinction between a description of the tumor which we make for our records and a diagnosis. The diagnosis is the basis for clinical work. I think only those features which have a relevance for the prognosis or for the treatment of the patient should actually be mentioned in the classification.

My own hypothesis has been that if such a primitive tumor contained some areas of higher differentiation, then probably the prognosis would be better. We tried to find out from our material on 62 of these tumors whether or not this is correct. I must say at the present time that we could not establish a better prognosis if we found areas of astrocytic or neuroblastic differentiation.

We have just a single case of a child with a rather undifferentiated tumor, operated upon at the age of eight months, who turned up two years later with recurrent tumor which was highly differentiated. We called it a ganglioglioma. This child is alive eight years later, but in all the other cases the prognosis was as bad as in those with undifferentiated tumors. I recall that two years ago, we discussed this problem with Dr. Rorke and she told us that, in her experience, the more differentiated tumors had an even worse prognosis than the highly undifferentiated tumors. We decided that it might be that when the tumor is more differentiated, then its susceptibility to radiation or chemotherapy is reduced. In such cases a recurrence may occur earlier, and the prognosis may be worse than in cases with completely undifferentiated tumors.

Therefore, before we can put these very interesting views into a classification, we should probably work together to get more material and find out what features are really relevant. The work of Dr. Gilles is a good basis for such research. He told us it is not yet appropriate for final conclusions, but it is at least a very interesting approach using some histologic characteristics. By using morphometry, we have now the tools to also do automated microscopic image analysis of different characteristics of the cells. So, I think we should continue on this path. Perhaps in another year or two when we again discuss the Blue Book, it might be that the time will be ripe for changing the classification. But for the time being, it is premature to divide the primitive neuroectodermal tumors into many subgroups according to these areas of differentiation.

DR. ISHIDA: We have studied 16 medulloblastomas using polyclonal antibodies raised against S-100 protein, GFAP and NFP. In 13 cases, GFAP and S-100 protein-positive cells were found irregularly scattered within the tumor tissue. On the basis of their size and shape they are mostly interpreted as *reactive astrocytes*. Three cases, however, contained GFAP-positive small cell forms, and also small numbers of cell forms immunoreactive for 68 Kd and 160 NFP.

With electron microscopy, microtubules and cell junctions of intermediate type were almost invariably found in cells of medulloblastoma. Within the limits of our observations, there was no convincing evidence indicating differentiation along either neuronal and glial cell lineage except for one case in which the cell processes contained clear and dense-cored vesicles.

In our observations the results of immunohistochemistry and those with electron microscopy appear conflicting. I have some doubts as to what extent the immunohistochemistry is to be considered as a reliable indicator for cell differentiation when applied to the study of immature neoplasms.

DR. KLEIHUES: I would like to make a few comments on the data presented by Dr. Darell Bigner, particularly on the activity of alkyl transferase. I think his data, and I won't criticize him because some of them are from our laboratory, showed quite convincingly that not only for the chemotherapeutic agents, BCNU and CCNU, but also for simple agents like procarbazine, and so on, the activity of these enzymes can be very important for the outcome of chemotherapy.

However, in my view, a word of caution is necessary. The simple fact is that there are Mer minus and Mer plus cell-line tumors, but there are no Mer minus tumors. When you take a tumor directly from the operating theater, and determine biochemically the 0-6, alpha 1 and alkyl transferase, all tumors contain the enzyme. We have looked at more than 20 tumors, and all of them contain alkyl transferase. It is only after you bring these cells to culture and develop lines, that you will find negative ones and positive ones. Then when you transfer them into nude mice, you find a good correlation with the outcome of chemotherapy.

If Dr. Bigner succeeds in developing an antibody which can be used in sections, we may learn more about the heterogeneity within the tumor, but Mer minus tumors in our experience do not exist.

DR. BIGNER: You cannot determine with the brine measure assay for measuring endothelial cells whether or not they exist. There are all sorts of normal cells in these tumors and until you can segregate them with an antibody or some cytologic method, you do not know which cells are Mer minus or Mer positive. In any tumor there is a large proportion of inflammatory cells as well. This is where I think antibody technology is going to need to be used in parallel with enzyme measurements and adjunct measurements. Nevertheless, I remain very hopeful about this whole approach. I think that we are going to be forced to look at these things.

DR. KLEIHUES: Yes, I agree about the intratumoral heterogeneity of the distribution of this enzyme. Absolutely nothing is known. However, it was present in fairly high amounts in the normal brain, in all tumors investigated, and was highest in benign tumors, like meningioma and neurinoma and the dedifferentiated astrocytomas. Therefore, I would anticipate that even with immunohistochemical techniques Mer minus tumors will not be found.

CHAPTER VI

Pinealomas: Need for an Ultrastructural Diagnosis

JACQUES HASSOUN and Danielle Gambarelli, *Laboratoire d'Anatomie Pathologique et de Neuropathologie, Faculté de Médicine, et l'Hopital de la Timone, Marseille, France*

The pineal region is a highly privileged area of the brain. It includes many cell types of glial or non-glial origin. The pineal body itself contains a wide range of cells, pinealocytes, astrocytes, oligodendrocytes, ependymocytes, ganglion cells and even Schwann cells from the conary nerve. Moreover, the high frequency of germinomas, suggests that germ cells are another possible cell type of the human pineal body. This explains why a variety of tumors can develop in this region. The most accurate means of diagnosis, in this area, remains the stereotactic biopsy but the main disadvantage of this method is the small size of the biopsied specimen. It is the reason why electron microscopy appears, even more than immunohistochemistry, to be a useful tool for diagnosis in the frequent cases of small round cell tumor. Pinealomas offer a typical example of such a diagnostic process.

Of the ten pinealomas collected in our neuropathology department since 1978, eight were in females and two in males. The age varied from 3 to 72 years, with only three children. A stereotactic biopsy was made in each case. The treatment in all cases was surgery followed by radiotherapy or chemotherapy. The course was rapidly fatal in five cases with occurence of spinal cord metastases.

Widely accepted classifications have proposed two types of pineal parenchymal tumors, immature pineoblastomas and more differentiated pineocytomas (RUSSELL and RUBINSTEIN 1963; RUBINSTEIN 1981b). In a study of 28 cases, HERRICK and RUBINSTEIN (1979) isolated several subgroups within each type according to the cytologic maturation of the tumor cells towards photoreceptor, astrocytic and/or neuronal cells. By light microscopy, our personal cases were difficult to place in one these precise subgroups. They. all presented transitional patterns between the Homer Wright rosettes of pineoblastomas and the larger fibrillary rosettes of pineocytomas. The lobular architecture described in some pineocytomas was never seen, no more than the mosaic pattern of pineoblastomas evoking the fetal pineal gland. So the less precise term of pinealoma appeared more convenient. As far as we are aware, ultrastructural studies of human pinealomas are very rare in the literature. Few specific works have been published until now (NIELSEN and WILSON 1975; HERRICK and RUBINSTEIN 1979; KLINE et al. 1979). The present study of ten tumors has permitted the discovery of several constant ultrastructural features which can be considered as specific markers for human tumor pinealocytes.

The first striking finding corresponded to the presence of dark and clear tumor cells joined with extensive *zonulae adherentes* and lying among pleiomorphic processes. They appeared as two variants of tumor pinealocytes. However, dark cells specifically showed microvilli and/or short tapered processes. Another distinctive pattern found only in five cases was represented by concentric arrangements of both types of cells and of joined processes around vacuoles. Such a complex intricateness of tumor cells ruled out the possibility of an edematous distension of the smooth reticulum and of the extracellular space. It could be related to the vacuolar system described in pinealocytes of some mammals (PEVET 1979; 1981). Calcareous concretions originate within these vacuoles (WELSH and REITER 1978).

Dense-core vesicles (DCV) have already been mentioned (NIELSEN and WILSON 1975) or illustrated (HERRICK and RUBINSTEIN 1979; MARKESBERY et al. 1981) in human pinealomas. They were herein located within cell perikarya and processes and associated with clear synaptic-like vesicles and multivesicular bodies only in the processes or bulbous endings. HERRICK and RUBINSTEIN (1979) supported the neuronal differentiation of tumor pinealocyte. However, they have been described in the pinealocytes of practically all the mammalian species studied to date (PEVET 1979; 1981) but not mentioned in human fetal pineal gland (MØLLER 1974). Like DCV, the clear vesicles, numerous in the present cases, appear as a normal component of the mammalian pinealocyte (PEVET 1981). Vesicle-crowned rodlets (VCR), also called *synaptic ribbons* have been observed in eight of our cases. These minute organelles were difficult to find, needing an appropriate investigation. They have been described in pinealocytes of most of mammalian species, including primates (PEVET 1979) and reported in tumor cell processes in a pinealoma (HERRICK and RUBINSTEIN 1979) and in a case of medulloblastoma studied *in vitro* (HERMAN and RUBINSTEIN 1984). The physiologic significance of these organelles is debatable. In mammals they are considered as phylogenetic remnants of lower vertebrate pinealocytes where they play a role as photoreceptors (WURTMAN et al. 1968). It is nowadays accepted that they do not play a true synaptic role in mammals and that their number depends on environmental lighting conditions (VOLLRATH 1973; KARASEK and VOLLRATH 1982), sexual hormonal factors (KARASEK and VOLLRATH 1982) and adrenergic innervation (KING and DOUGHERTY 1982). The vesicular structures related to the VCR were similar to the *synaptic* spherules (KARASEK and VOLLRATH 1982; KING and DOUGHERTY 1982; HEWING 1980) VCR and spherules are therefore logically to be expected in human normal and tumor pinealocytes, without having the significance of a neuronal differentiation. Contrariwise, normal or abnormal synaptic complexes described in true neuronal tumors of the brain (RUBINSTEIN and HERMAN 1972, HASSOUN et al. 1982) and of the cerebellum (SHIN et al. 1978) were not found in the present tumors. Micro-tubular sheeves (MTS), observed in four cases, have never been described in human pineal tumors but have been reported in papova-virus induced pineocytomas of the Syrian hamster (VARAKIS and ZU RHEIN 1976). Originally described in the adult rat pinealocyte (WOLFE 1965), they appear as a unique feature in some species and probably derive from centrioles during cell differentiation. In our material their presence, close to these organelles, highly favors this hypothesis. MTS might be transformed into synaptic ribbons (RUIZ-NAVARRO et al. 1982). Such a spatial correlation could not have been made in the present cases. Paired twisted filaments (PTF) forming helices were found in the perikarya of tumor cells in seven cases. The filaments were about 8 nm wide. The helice had a 26-30 nm periodicity. Totally different from Alzheimer neurofibrillary tangles, these filaments could be a specific feature of human tumor pinealocytes. Other structures observed herein have been characteristically described in mammalian pinealocytes and found in some human pinealomas: membranous whorls and annulate lamellae (KLINE et al. 1979; PEVET 1979), glycogen granules (NIELSEN and WILSON 1975; ROMIJN et al. 1976; MARKESBERY et al. 1981). In human tumors, some authors (NIELSEN and WILSON 1975; KLINE et al. 1979; MARKESBERY et al. 1981) stressed the presence of cilia also seen in great number in four present cases. They have been described in human fetal pineal gland (MØLLER 1974) and constitute a typical but nonspecific organelle. Heterogeneous cytoplasmic inclusions were particularly numerous in our cases. Identical structures were observed by KLINE et al. (1979) who compared them to granular bodies of fetal human pinealocytes (MØLLER 1974). Nevertheless, this author described homogeneous but not crystalline or fibrillary inclusions. Most of the inclusions described here also appeared different from lipofuscin and neuromelanin granules of human substantia nigra and locus coeruleus (HIRANO 1971). The significance of these peculiar cytoplasmic inclusions remains to be elucidated. Numerous fibrous bodies were observed in three of the present cases and appeared as a characteristic feature of these tumors. To our knowledge, they were neither reported in mammalian pineal gland nor in human pinealomas. They did not bear any resemblance to astrocytic microfilaments otherwise observed in the glial processes of the tumor stroma. The very rare GFAP-positive glial cells or processes found in all our cases did not permit any more than the ultrastructural findings to retain the possibility of an astrocytic differentiation from the tumor pineal cells. GFAP-positive cells have been reported in rat (MøLLER et al. 1978) and in human pineal gland (LOEWENTHAL et al. 1982) and are to be considered as a normal component of this endocrine tissue. S-antigen immunoreactivity has been described as a possible marker of pinealocytes in tumor processes (KORF et al. 1986; PERENTES et al. 1986).

To conclude, it can be asserted that, whatever the histologic aspect of the pineal parenchymal tumors (i.e. pineoblastoma or mixed pineoblastoma-pineocytoma), numerous ultrastructural features constantly evoke those of mammalian pinealocytes (VOLLRATH 1981). The association of intermingled dark and clear cells, intra- or extra-cellular vacuolar spaces, pleiomorphous cell processes and the presence of some cytoplasmic organelles (DCV, clear vesicles, VCR, MTS, microtubules, membranous whorls, PTF, fibrous bodies, heterogeneous cytoplasmic inclusions) give to these tumors a very specific pattern which can be distinguished from that of the neuronal and glial tumors developed in the pineal area.

Acknowledgement: We are grateful to Mrs. E. Pasquale, F. Fabre and Mr. R. Gochgagarian for their technical assistance.

Discussion of Chapter VI

DR. ZÜLCH: The WHO classification (ZÜLCH 1979) distinguishes between the *pineocytoma*, which is a rare isomorphous tumor with uniform cytology and processes radiating toward the vascular stroma (grades I–III), and the *pineoblastoma*, which is also a rare and highly cellular, malignant pineal neoplasm, very closely resembling the medulloblastoma (grade IV).

Since we have under consideration the classification as published in the Blue Book, I want to say a few words about pineal tumors. We have had two different approaches to terminology for some tumor groups. This has been the case with the pinealomas, about which we have just heard a very interesting analysis.

In Europe, the traditional way of classifying is to distinguish between the pinealoma (the *two-cell type* tumor) and its malignant counterpart. In the United States, the concept is prevalent that the large majority of tumors of the pineal gland are actually *germinomas*. The same is true for the *ectopic* two-cell pattern tumor near the chiasm. My belief, supported by personal investigation, is that there are two entities: (a) a real tumor of the large cells of the pineal parenchyma showing the typical processes with club-like expansions at their tips, as described by Del Rio Hortega (see CLARA 1959), intermingled with *lymphoid* cells (formerly called *anisomorphous pinealoma* in our instition), and (b) the *germinoma*, a common tumor at this site, which does not originate in the pineal epithelium.

At the last moment in our earlier deliberations, this was changed, and we accepted a condensed classification into *pinealocytoma* or *pinealoma*, which are essentially the same, and the terms *pinealoblastoma* and *medulloblastoma of the pinealis* were not accepted. Later the majority came to the conclusion that so-called *two-cell* tumors are two-cell type germinomas and that there are no pinealomas of the two-cell type. Therefore, we changed the twofold reclassification of the tumors of pineal pattern at the last moment, and you will find in the Blue Book the two-cell tumors termed only as *germinomas*.

KRABBE (1923) called the two-cell type tumor a *pinealoma*. Somewhat later, we used the term *pinealocytoma*. Finally, FRIEDMAN (1947) came up with the statement that the similarity between tumors of the testis, the seminomas, and the two-cell tumors of the pineal was so close that we must accept those as real germinomas. Since then, in the English-speaking world, the term *germinoma* has been prevalent. However, the *pinealoma* has maintained its importance for a long time in Europe.

I have always fought personally for the duality of this tumor type and gave in only at the last moment. We had always kept in mind that we were a very democratic group with the majority view being accepted. Yet I want to support here my personal view since I still adhere to this duality. I am supported in this position by the prominent senior neurosurgeon of Japan, Professor Keiji Sano of The Imperial University (SANO 1983, SANO 1985). The Japanese have had a greater experience in the field of pineal tumors than we have in the West because in the Eastern countries, China, Japan, and Korea, the incidence of pinealomas of the two-cell type is about ten times as high. However, the cause of the preponderance in the yellow race is unknown.

There is a difference in the histologic appearance of the two-cell type pinealoma and that of germinomas. The latter frequently contains large, sometimes multinucleated cells, and the two-cell pattern is usually not as prominent.

In our material the pinealomas were positively impregnated by the De Girolami method and they were PAS negative, which means that there were no mucous substances. The opposite was true for the germinomas. They were PAS positive and De Girolami negative. Sano reported some even more convincing evidence regarding differences in the architectural pattern.

Moreover, in Japan, MATSUTANI et al (1975) and MATSUTANI (1976) proved immunologically that the large cells of the two-cell type tumors were derivatives of pineal parenchymal cells as BARGMANN (1943) had previously described. Finally, in the spinal fluid of patients with pinealomas one can find the same pattern of parenchymal cells as was described originally for the pineal gland by Del Rio Hortega. GLOBUS and SILBERT (1931), who wrote a very early paper on that problem, drew attention to the similarity between pineal tumor cell pattern and the stages of the developing pineal gland. Furthermore, the small lymphoid cells are, according to KALM and MAGUN (1950) who made an important study using large, celloidin section from such a case with metastases, immature tumor cells. This is a concept shared originally by Dorothy RUSSELL (1944) and by FRIEDMAN (1947) who called them *small maturing cells*.

Pineal ectopic tumors can be, but are not necessarily, primary tumors. Observations definitely show a seeding of pineal tumors, via the cerebrospinal fluid pathway into the region of the pituitary stalk. I have seen several cases at autopsy, and the case of KALM and MAGUN (1950) is of that type.

Pineal tumors, even of the so-called *germinoma* type, are frequently heavily calcified. General pathologists have told me that the testicular germinomas are not. At the International Neurosurgical Congress in Toronto, Sano emphasized again the dualistic concept and stated his opinion that 20% of the ectopic type were also of true pineal origin. We will have to wait for more modern methods to further clarify this matter.

DR. KEPES: Like everyone else in this room, we have looked at many pineal tumors. Most of us have grown up with the original BAILEY and CUSHING classification that described the two-cell pinealoma as one that is a replica of the developing, fetal pineal gland. Dorothy RUSSELL (1944) observed the great similarity between these mixed-cell pineal tumors and the germinomas that she, as a general pathologist, had seen in other parts of the body, notably the testes and ovaries. She introduced a concept that would be very hard to refute even today. If I may, I would like to mention a few arguments, in addition to the ones that are already known to us, that would support this notion of germinomas.

One of them is that true pinealocytomas, the kind that have photoreceptors identifiable by electron microscopy, are seen in the pineal region, whereas to my knowledge, they do not occur in the suprasellar area, where, for example, the double-cell tumors are very common.

Second, when we talk about the two-cell type, we should not overlook the point that there are more than two cells. A number of these tumors, in addition to containing small lymphocytes, have histiocytes and multinucleated giant cells, so that a whole host of cells, some even showing a granulomatous reaction, may be present.

Third, lymphocytes can be identified as lymphocytes, and finally, and very importantly, something you would not expect to see in true pineocytic tumors, is the further development toward embryonal carcinomas, entodermal sinus tumors, and choriocarcinomas. All of these changes have been observed in these locations, and I don't think the true pinealocytomas would do that. More specifically, choriocarcinomas in pregnant women are much more common in the Orient, where there is a general tendency toward increased numbers of germ cell tumors, than in Europe.

DR. ZÜLCH: You characterize very well the germinomas and I would never object. I said 20% are true pineal parenchymal cell tumors, and 80% may be germinomas. Like Sano, I have had a dualistic concept as I have already indicated. That is what I have been fighting for.

DR. KLEIHUES: But where is the proof? We have followed your presentation with great interest but it was mainly historical. I think there is now, particularly for the germ cell tumors, considerable information regarding specific markers. We have for the germinomatous portion the placenta, alkaline phosphatase, alpha-feto protein and beta HCG. All of these markers for germinal tumors are present in various compositions and in varying amounts.

So, if you propose that a two-cell type tumor exists which originates from the pineal gland and which is not a germ cell tumor, I think you should give us these tumors so that we can see whether they contain beta HCG or alpha-feto protein. If they do not contain these markers, we will talk about it again. I'm sorry but I do not think at the present time you have any evidence.

DR. ZÜLCH: How many germ cell or two-cell tumors have you examined with 100% of these markers?

DR. KLEIHUES: They rarely have 100% but the germinomatous part, in all cases that we have investigated, has placenta alkaline phosphatase. Moreover, Dr. Scheithauer in a large series of germinal tumors from the Mayo Clinic, found a rather large proportion, about 8 to 3 as far as I can recall.

The term *pinealoma* is rather general. Some people use it for tumors originating from the pineal gland, and usually exclude the germinomas. Other people call any tumor which arises in the pineal gland a *pinealoma* and use it as a kind of geographic term. Dr. Hassoun used it more specifically, if I understood correctly, for tumors of known germinomatous origin which are residing in the pineal gland. Is that correct?

DR. HASSOUN: That's correct. I would like to answer Prof. Zülch briefly by referring again to our series of germinomas. We observed 22 germinomas by electron microscopy and always found the same picture and the same features. Germinoma cells are round and absolutely devoid of any processes. They are absolutely different from neuroectodermal cells. Their cytoplasm is filled with a large amount of glycogen, which explains the pH positivity seen by light microscopy.

Another very specific feature is an active inflammatory reaction with large macrophages and also lymphocytes. Often it is possible to find as many macrophages as tumor cells in some fields. This feature seemed to us very specific for this kind of tumor since the macrophages showed these distinct features of phagocytosis of tumor cells, which were recognizable, thanks to the glycogen particles. The brain tumors have these cells, but macrophages are also found in other germinomas of the body, such as those arising in the mediastinum, the testes, and ovaries. These very special tumors are completely different from the true pinealomas.

DR. ISHIDA: In Japan, germinomas comprise about 3-4% of primary brain tumors. Recently, *Neurosurgical Journal*, published in China, reported statistics of surgically removed brain tumors based on the study of more than 1900 cases. In these statistics, pinealoma (probable germinoma) comprises 3%. The incidence of germinoma in China is also as high as it is in Japan.

MATSUTANI et al. (1975) reported the presence of germinoma-like pineocytoma. However, most Japanese pathologists and neuropathologists would not believe in the existence of such tumors because they lack evidence of cell differentiation along pineal cell lineage.

General Discussion of Part One

DR. BARNARD: One can attempt to summarize the numerous papers we have heard; or one can select small pieces, or nuggets of information, which are of value to all of us; or one can attempt in these circumstances to be anecdotal and try to put the contents of the day's session into a wider context. I shall probably attempt to do all three, and it is unlikely that I shall succeed.

The papers that we have heard on various topics have in common certain fundamental themes. One is the difficulty of finding accurate widely acceptable terminology, and this is especially the case in respect to the undifferentiated embryonal or *primitive* tumors of the central nervous system. This difficulty has been recognized since the time when classification of the nervous system tumors began.

BAILEY and CUSHING (1926) were innovators, and as such, they were able to collect material and propose names for different tumor entities, correlating morphologic appearance with clinical outcome. Their work was more than influential; it was fundamental to the cytogenetic basis of classification, using the resemblance of tumor cells to their counterparts in development. Many of the terms which they pioneered in their day have become more clearly defined as knowledge has accumulated, but difficulties over the application of certain terms and their precise definitions have remained.

DEL RIO HORTEGA (1933) proposed a scheme of classification which perhaps went further on a cytogenetic basis than any other, inasmuch as it only related to the precise cellular composition without any clinical implications. Subsequent authors have laid more emphasis on the clinical implications, and have stressed the need for simplicity in application. These considerations were well known to the WHO investigators who set about their task of compiling a classification to include all the well-known diagnostic entities. Above all, the task was to facilitate communication between pathologists of different nationalities, coming from different backgrounds, and employing different investigational methods. This work really went through the decade of the 1970s, and the Blue Book, which was produced in 1979, has several good features (ZÜLCH 1979). I think we would agree that it is concise, easy to read, well illustrated, and above all, it is multilingual. It is satisfactory in many ways that such a useful classification could be produced, for it resulted in part from a spirit of compromise and also in part because it was based on the principle of giving alternative names to identify the same tumor.

Histological Classification of Tumours of the Central Nervous System (ZÜLCH 1979)

I. Tumours of Neuroepithelial Tissue
  A. Astrocytic Tumours
    1. Astrocytoma
      a. Fibrillary
      b. Protoplasmic
      c. Gemistocytic
    2. Pilocytic astrocytoma
    3. Subependymal giant cell astrocytoma (ventricular tumour of tuberous sclerosis)
    4. Astroblastoma
    5. Anaplastic (malignant) astrocytoma
  B. Oligodendroglial Tumours
    1. Oligodendroglioma
    2. Mixed oligo-astrocytoma
    3. Anaplastic (malignant) oligodendroglioma
  C. Ependymal and Choroid Plexus Tumours
    1. Ependymoma
      Variants:
      a. Myxopapillary ependymoma
      b. Papillary ependymoma
      c. Subependymoma
    2. Anaplastic (malignant) ependymoma
    3. Choroid plexus papilloma
    4. Anaplastic (malignant) choroid plexus papilloma
  D. Pineal Cell Tumours
    1. Pineocytoma (pinealocytoma)
    2. Pineoblastoma (pinealoblastoma)
  E. Neuronal Tumours
    1. Gangliocytoma
    2. Ganglioglioma
    3. Ganglioneuroblastoma
    4. Anaplastic (malignant) gangliocytoma and ganglioglioma
    5. Neuroblastoma
  F. Poorly Differentiated and Embryonal Tumours
    1. Glioblastoma
      Variants:
      a. Glioblastoma with sarcomatous component (mixed glioblastoma and sarcoma)
      b. Giant cell glioblastoma
    2. Medulloblastoma
      Variants:
      a. Desmoplastic medulloblastoma
      b. Medullomyoblastoma
    3. Medulloepithelioma
    4. Primitive polar spongioblastoma
    5. Gliomatosis cerebri

Professor Zülch has collected information on the reception of the book from different types of user. The majority regarded it as good, even sometimes very good. In the text a numerical grading scheme was used and the value was based largely on the experience of each tumor type and its level of differentiation.

There are different stages of classification. First of all, taking the tumors of neuroepithelial tissue, the astrocytic group is given an overall "A", and then the astrocytomas, the different types are given numerals 1, 2, 3, 4, 5. You will comment, of course, on the omission of a pleomorphic xanthoastrocytoma, which at that time, had not achieved its separate status. We felt very much that it was possible to diagnose a tumor in terms of a level of classification, for instance, if one could only say it was an oligodendroglial tumor, not otherwise specified, one could only give it a "B" rather than a "B-2" for a mixed oligoastrocytoma.

The principal criticisms that have been voiced, some of them eloquently expressed, have been largely on the application of this scheme to childhood brain tumors. The proposed revision, with which Drs. Rorke, Gilles, and Davis were associated (Rorke et al. 1985), presents basically a scheme for the astrocytic, oligodendroglial, and ependymal tumors, very similar to the WHO classification, with a greater degree of emphasis on the mixed tumors which are subdivided as shown below. Glioblastomas are placed in a separate category.

Tumors of Neuroepithelial Tissue (Rorke et al. 1985)

I. Tumors of Neuroepithelial Tissue
   A. Glial tumors
      1. Astrocytic tumors
         a. Astrocytoma (fibrillary, protoplasmic, gemistocytic, pilocytic and xanthomatous)
         b. Anaplastic astrocytoma
         c. Subependymal giant cell tumor (tuberous sclerosis)
         d. Gigantocellular glioma
      2. Oligodendroglial tumors
         a. Oligodendroglioma
         b. Anaplastic oligodendroglioma
      3. Ependymal tumors
         a. Ependymoma
         b. Anaplastic ependymoma
         c. Myxopapillary ependymoma

4. Choroid plexus tumors
    a. Choroid plexus papilloma
    b. Anaplastic choroid plexus tumor (carcinoma)
5. Mixed gliomas
    a. Oligoastrocytoma
    b. Ependymoastrocytoma
    c. Oligoastroependymoma
    d. Oligoependymoma
    e. Subependymoma - subependymal glomerate astrocytoma
    f. Gliofibroma
6. Glioblastomatous tumors
    a. Glioblastoma multiforme
    b. Giant cell glioblastoma
    e. Glioblastoma with a sarcomatous component
7. Gliomatosis cerebri

B. Neuronal tumors
1. Gangliocytoma
2. Anaplastic gangliocytoma
3. Gangliogloma
4. Anaplastic gangliocytoma

C. *Primitive* neuroepithelial tumors
1. *Primitive* neuroectodermal tumor, not otherwise specified (NOS)
2. *Primitive* neuroectodermal tumor with:
    a. Astrocytes
    b. Oligodendrocytes
    c. Ependymal cells
    d. Neuronal cells
    e. Other (melanocytic, mesenchymal)
    f. Mixed cellular elements
3. Medulloepithelioma
    a. Medulloepithelioma, NOS
    b. Medulloepithelioma with:
        (1) Astrocytes
        (2) Oligodendrocytes
        (3) Neuronal cells
        (5) Other (melanocytic, mesenchymal)
        (6) Mixed cellular elements

D. Pineal cell tumors
1. *Primitive* neuroectodermal tumor (see C above) (pineoblastoma)
2. Pineocytoma

The WHO scheme, listed previously, shows the pineal group, the neuronal group, and group "F", the principal source of controversy that we have been hearing about today.

Many people using this scheme in a practical way found that glioblastoma, shown as "F-1", did not assort very well with the others in this group of embryonal neoplasms and felt that it should be removed from this category and placed in a separate one.

Another possible improvement would be to amalgamate the remaining items in "F" with some of those in "E". In other words, assuming that many medulloblastomas are in essence of neuroblastic type, they would be more appropriately grouped together with neuronal tumors. The primitive polar spongioblastoma, of course, is another matter, and that might well have to move into another category such as the astrocytic group. This kind of fine tuning would not present any fundamental problems.

The pineal tumors, referred to in the classification, are the sort of tumors which Dr. Hassoun talked about. The question of the two-cell type pinealoma was decided by a majority of the WHO participants accepting that these entities could not be identified as true pineal tumors, and that the germinoma can be found elsewhere (under germ cell tumors).

Today we have heard a great deal about the way in which the poorly differentiated and embryonal tumor group could be reclassified. Dr. Sobin reminded us that it was important to separate entities from concepts, and that, in fact, the most helpful thing we can do is to remember that *entities* are specific diagnostic categories based on a morphologic description.

Dr. Rorke took us very lucidly over the concepts underlying the primitive and the embryonal group of tumors, and discussed the various possibilities regarding the neural tube and its derivatives, and the potential of the external granular layer cells. As she said, the pathologist is not a prophet, and the question of how far one can go on the basis of H&E, or H&E plus GFAP and neurofilament, in describing these primitive tumors is, of course, a controversial issue. The cerebellum, she told us, does not contain a unique bipotential cell, and the primitive neuroectodermal tumor may arise anywhere in the nervous system.

Dr. Rubinstein pointed out that in any orderly scheme, existing defined entities must be incorporated and the scheme must be consistent; it must not cause confusion in the mind of the non-specialist.

Now I return to the classification which was proposed as a modification of the WHO by Rorke et al. (1985), showing the *primitive* neuroepithelial tumors, some containing astrocytic, oligodendrocytic, and ependymal cells. The *primitive* neuroectodermal tumor of the pineal or pineoblastoma is included in the pineal cell tumors.

Dr. Rubinstein stressed that any classification must have clinical relevance. From a practical point of view, he considered that putting PNETs into one category does not help since it results in the possibility of incorrect diagnoses by the non-specialist. It underplays the fact that localization of tumors in the nervous system is important. He went on to demonstrate that neoplastic disease, which essentially involves cells of renewal, depends on the concept of the window of neoplastic vulnerability. In the cerebellum, different from the forebrain, mitotic division continues through the first postnatal year.

He went on to describe a special entity: the desmoplastic ganglioglioma of infancy, a distinctive tumor worth recognizing because its prognosis is essentially a different one from many of the poorly differentiated and embryonal tumor group.

In discussion a lot of interesting points were raised on these two papers. For instance, Dr. Kleihues pointed out that genetic differences are important in the evaluation of retinoblastoma.

One point I would take from Dr. Bigner's presentation concerns pathology as a discipline in the future. He stressed that pure morphology on light microscopy is becoming inadequate; a new generation of monoclonals and enzymes were coming into being, requiring fresh tissue. Retrospective studies using paraffin-embedded material will be less valuable.

Now, I think I should touch upon the alternative strategy proposed by Dr. Gilles, in which he examined groups and clusters of prognostic features, which he analyzed in respect of posterior fossa tumors independent of a specific pathologic diagnosis. Some of us were slightly skeptical of the value of this method without reference to the pathologic diagnosis, but I think we all value very much the information that was given us.

Dr. Becker talked most interestingly about the levels of diagnosis which are possible, depending on the techniques available to any particular laboratory. The ideal may be impossible for many people, though we would all like to come further down his numerical scale. He also emphasized the spectrum of variation for the use of the term *primitive neuroectodermal tumor*, and pointed out that *primitive* in this context really does not necessarily imply embryonal. I think this is of great importance because we are faced with the practical point that some people are using this term to mean adult tumors, which are probably anaplastic rather than embryonal type.

What is remarkable is that in 1988 we still continue to talk about *small tree glue*, for that is what oligodendroglia means. Is it not also remarkable that this word has acquired a precise secondary meaning, so that when any of us uses it, it immediately brings to mind a very clear picture. During my lifetime, the whole of the study of pathology seems to me bedeviled by lack of agreement over the use of words. One of my early mentors, to be anecdotal, objected to my using the term *necrosis* for death of tissue. Necron, he said, is a corpse. Thanatos is death, so *thanatosis* would be more appropriate. This has not caught on.

While my mentor, Professor McMenemy, was unwell, I helped Professor Symmers and Professor Payling Wright with their monumental book on pathology, In the section on neuropathology, Professor Symmers refused point blank to accept the term *perivascular* for the Virchow-Robin space since he said it should be *circumvascular*. He also refused to accept *microglia* since the microglial cell is not a glial cell, and the term which he favored for cerebral lymphoma, as we now know it (then known as the *perithelial sarcoma* or *microglioma*), was the *circumvascular Hortega-cell sarcoma*, which on the whole has died an unlamented death following the atrophy of disuse.

In conclusion: in 1979, under the inspired leadership of Professor Zülch, a classification was eventually produced (ZÜLCH 1979), uniting the views of many different schools in a consensus, the way that we looked at a large number of sections and reported them independently. The one thing that united us as a group was the belief that language must be used as an effective tool to convey meaning with as near as possible a universal application. The thing that we now have to discuss is whether this classification can be fine-tuned to take note of existing improvements and new entities (RORKE et al. 1985) which have come to light in the last few years, or whether a more radical realignment has to take place in order that the system can be brought up to date and made more useful.

DR. EARLE: We have been formal, and I think there is a great virtue in formality up to a point. However, I would like to reminisce a little bit and tell you about the problems that I have had in reaching a consensus on the classification of brain tumors, either in my own mind, or in consort with my colleagues. Back in the early forties, when I graduated from medical school, there was only one classification, and it was that of Bailey and Cushing.

Then after some time in the Navy, I decided to go to Montreal to be a neurosurgeon, and wound up as a neuropathologist. I was shocked the very first week or two to find out that meningiomas did not exist in Montreal; they were *meningeal fibroblastomas*. There was no such thing as a nerve sheath tumor; it was a *perineural fibroblastoma*. There was some kind of astrocytoma which was called *gemistocytic*, a fat astrocyte, and there was one called *pilocytic*, which was explained in detail to mean that the processes appeared "as hairs waving in the breeze". The Penfield classification was popular in many countries but the Bailey and Cushing classification was the most popular in the United States.

So, finally, having somewhat adjusted to this, I was brought out of my dilemma by George Olszewski, who was translating into English a small book by Professor Klaus Zülch. Most likely he would have been thrown out of the Montreal Neurological Institute if he had popularized it too greatly, but, at least, he managed to keep down the number of terms used. However, there was more consensus between Penfield and Zülch than would immediately meet the eye.

Soon afterward, I became aware that the Armed Forces Institute of Pathology, formerly the Army Medical Museum, had come out with some fascicles that were rather cheap, so that most of us could afford them. The first series on brain tumors was written by Kernohan and Sayre. This, of course, was very much embraced by the neurosurgeons because it had a lot of simplifications in it. Astrocytomas were only of four kinds, grades I, II, III, and IV and all of these other names to which we had been exposed did not reach a consensus among neurosurgeons, I can assure you. This new fascicle had a very great impact. I don't know how many thousands of those volumes were printed, but in traveling around the world, I have yet to go into a pathology department and not be able to find one.

Then the nerve sheath tumor fascicle by Arthur Purdy Stout came out, in which he introduced the term *neurilemoma*, and he spelled it L/E/M/O/M/A, one "M", for which he was criticized by all the Greeks in the world. They said, "everyone knows that lemma has two M's." But this, too, did not completely take hold. George Olszewski, like Klaus Zülch, consistently referred to the nerve sheath tumors as neurinomas. Bailey, on the other hand, used the term neuroma. The former spelling was more widely used, especially in Europe.

Arthur Purdy Stout's neurilemoma caught on for a very short time. Then, along came the electron microscope, which resulted in a great argument as to whether the cells were fibroblasts, as Penfield had maintained in promoting the term *perineural fibroblastoma*, or whether they were Schwann's cells. Finally the "Schwann's cellians" won, and so the term *schwannoma* became extremely popular. I found, however, in my travels around, that there was still no consensus as to whether we should call them one thing or the other.

Then, lo and behold, we all felt that the electron microscope was going to produce a consensus for us. It was going to show us all the wonderful detail inside the cells and we would no longer have to argue about these things. Unfortunately, that did not occur. In fact, Dr. Penfield was very reluctant to install an electron microscope. He felt that we had not yet exhausted the metallic impregnation methods. Then along came histochemistry. We said, "My goodness! instead of just tinctorial qualities based on pH, we are going to have a chemical reaction that will tell us the ultimate in classifying, and subclassifying tumors. It, too, did not work.

Finally, as Lauren Ackerman has said, "A brown haze descended over the field of pathology." This was called immunoperoxidase with all of the refined monoclonal antibody techniques. Again, we were confident that this was the end of our search.

A few years ago, at a meeting in Washington, we were all encouraged to become immunopathologists, or perish. Then at another meeting in Washington more recently, we were all told to either become molecular biologists, or at least become damn good friends with one.

Consequently, we are now in a very peculiar position before arriving at any consensus, in that it appears that neuropathology to some extent can't agree with itself. Recently 10 or 12 neuropathologists have split off to become a group of experimental neuropathologists and are going back to meet annually with the American Neurological Association. I used to go to those meetings in conjunction with the neurologists and neurosurgeons and found the exchange useful.

Now, with that having been said, let me tell you how I would like to begin with an attempt to arrive at a consensus in our discussion here. The Delphi procedure is widely used to do this. I had no idea, when I was engaged for the first time in a Delphi Procedure in Washington, what it was all about. It turned out that the NIH had several million dollars that they were going to spend to form cancer institutes, but they had to get a consensus on what an ideal cancer institute was, and where they were going to put all of those millions of dollars.

They had at least thirty people go to a hotel at Dulles Airport, where they had "professional consensus people", with reams of news print and pencils. It was planned that we would sit in groups and try to decide what are the characteristics of a superior cancer institute and what are our alternatives. This activity went on for three days. We were told originally that the Delphi Procedure which they were going to use was almost foolproof in arriving at a consensus. The Oracle of Delphi, we were told, sat upon a tripod with all of the vapors surrounding her and then the wise men sat around listening to what the Oracle said in strange languages, and eventually they arrived at a consensus. So the Delphi Procedure means that you are going to have to listen to the Oracle, and then you will arrive at a consensus of what you think has been said.

We have had a lot of Oracles of Delphi talk to us today. The first question we ought to answer is, "Do we want to change the nomenclature of pineal tumors?" Right now we have the pineocytoma, and the pineoblastoma. Is there a consensus among you that we should change these two terms?

DR. KEPES: I think the two terms are fine, but it has to be recognized that they are transitional terms.

DR. EARLE: You think it should be added then that they are transitional forms? Apparently no one disagrees with this so let us now turn to the germinoma. Does anyone want to put the germinoma back in and call it *pinealoma*, or *ectopic pinealoma*, or should it stay in the germinal series, or elsewhere?

DR. KLEIHUES: Let us leave it where it is.

DR. EARLE: Anyone else? Any disagreement?

DR. ZÜLCH: No, I was not for the germinomas in the first place.

DR. EARLE: Well, then, I think we have reached some agreement. I am amazed. I think that now we must get down to the "nitty-gritty".

What I would gather from talking to people regarding the terminology of primitive neuroectodermal tumors is that we have more of a consensus than may be immediately apparent. There is a considerable feeling, I believe, that we should adhere to the terms *medulloblastoma*, *pineoblastoma*, and *neuroblastoma* as they are in the Blue Book classification now. Is there any disagreement with that? Does anyone want to annihilate these terms?

DR. RUBINSTEIN: We might want to add one more. I think the medulloblastoma has now been sufficiently well authenticated by cases to deserve a mention on its own. Later, I hope to show some data justifying this term, and distinguishing it from malignant ependymoma, both on histologic grounds, but particularly on biologic and clinical grounds. I think that Dr. Becker agrees with the existence of pinealoblastoma and that this represents a true tumor entity, albeit a rare one.

DR. EARLE: Anybody disagree? Do you want to change that?

DR. RORKE: I would like to keep the terms in parentheses.

DR. EARLE: So there is another suggestion. Does anybody else want to keep it in parentheses?

DR. EARLE: Maybe we could leave *ependymoblastoma*, also *cerebral ependymoma*. There was a case presented at the Canadian meeting last year, in which an ependymoma independent of the ventricles, was shown. At least, under electronmicroscopy it had cilia and all the features of an ependymoma and it occurred in the cerebral hemisphere. We did not have that in the old classification.

DR. RUBINSTEIN: We had ependymomas. Can't they occur anywhere?

DR. EARLE: If we accept that, we should somehow make an explanation that these can occur independent of the ventricles. They have even occurred under the skin in the sacral region. We are all aware of that.

DR. RUBINSTEIN: We have heard many opinions and I think that each of us can fine tune them a bit. I am sure that Dr. Rorke, Dr. Becker, and Dr. Gilles also want to say something about the way they would like us to approach the differences. I think it is important to say something but we can't simply rush ahead on a consensus.

DR. EARLE: I agree wholeheartedly with that approach. I would preface this though by saying again that in reaching a consensus, it is necessary to start somewhere. In talking to people independently, there is considerable feeling that we should retain the terms *medulloblastoma, ependymoblastoma,* etc. The question to be resolved is "should we have a straight term for the primitive neuroectodermal tumor, which does not fit into anybody's classification, or should it simply be left as an unclassified tumor, as Kleihues has suggested?"

DR. RUBINSTEIN: May I sort of disagree? I think that's the crux of the matter. The difference in opinion is not very great but nevertheless, it is there. We have all been very polite and very nice to each other, but perhaps I might inject a little passion into the dialogue.

Even though the difference may not be very great, it is, I think, a significant one. Perhaps it may be a subjective one. The attempts by Dr. Gilles to be as objective as possible are very praiseworthy. Nevertheless,in our attempts to reach a diagnosis, there is inevitably a subjective element. We do get information on the site of the tumor, the age of the patient, the amount of tumor available, and all of these numerous variables that Dr. Becker has described so very well. All these variables come into the picture, so our diagnosis to some extent is bound to be a subjective exercise. Perhaps we ought to allow this to be so. We are entitled to have different diagnoses, different opinions. We certainly are entitled to change our minds. There are many things that I wrote in the fascicle, with which I no longer agree.

When it comes to the matter that Dr. Earle pointed out, the question of PNET, it happens that temperamentally, this is a subjective thing. We should have reasonably well defined entities before placing them in a classification. We ought to have entities well described, and then classify them, not the other way around. That is why I feel that at this stage we have not seen or clearly defined a cerebral medulloblastoma in man. We cannot state that it exists, nor can we make this concept a foundation for a classification.

In fact, my only quarrel with Dr. Rorke is that I cannot accept PNET as a diagnosis. I prefer to call these neoplasms *unclassified tumors*. The only entities I would recognize are the other ones that we mentioned before. That, I think, is the point of disagreement. I do not see how we can really come to a consensus on that in which two opposing parties hold very honestly, and very sincerely, such different views. In other words, I do not think that we can have a classification before we have well-defined entities.

DR. KEPES: Just coming to practical terms, I think the group of nonclassified or nonclassifiable tumors is a wider one than what we have understood about the primitive neuroectodermal tumor. I do not agree that nonclassifiable tumors in the brain, even though acknowledged to be primary, would necessarily be primitive neuroectodermal tumors.

DR. ZÜLCH: Well, I think we have to keep in mind what Dr. Sobin said in his introduction. The Blue Book is already available as a service book for the clinician. Neurosurgeons, radiologists, and others are quite well satisfied with it at the moment. We have a relatively simple book. If we now bring in new entities, such as medulloblastoma with a mixture of astrocytes, or six or seven types of gliomas, we haven't really provided any practical new information. We need much more material from various sources before we can say that this is something significant with respect to biologic behavior. We must still provide a service to the clinician.

DR. RORKE: I should like to repeat what I said earlier, and that is that the concept of primitive neuroectodermal tumor is not a diagnosis to be made by exclusion. After one is able to diagnose tumors in the cerebellum and call them *pinealoblastomas*, and also call tumors in the pineal *pinealoblastomas*, that one is then left with a category of tumor that is composed of these cells, but does not fit into these categories. That is not what I meant by my original concept of the primitive neuroectodermal tumor.

Dr. Rubinstein may have seen cases of tumors reported in the cerebrum. As a matter of fact, in the original group of cases reported by Bailey and Cushing (1925), there were, of course, cerebral examples, as noted also in Cushing's 1930 publication. He directs attention to these, and says that Bailey reviewed them, and they were still left with two examples that essentially looked like medulloblastoma, but they were in the cerebrum.

In his later writings, he points out that such tumors do occur in the cerebral hemispheres, and I think those of us who look at a large number of childhood brain tumors have seen tumors in the cerebrum that do not have the specific characteristics of what is called *cerebral neuroblastoma*, but have a variety of characteristics, which are really best described by some other designation than *cerebral medulloblastoma.*

The concept that I have tried to introduce, and which unfortunately has been misinterpreted, is not a *waste basket* category. It is a concept which says that in the central nervous system, primarily in children but in individuals of any age, you may have tumors which have a certain histologic appearance. The literature is very clear about the fact that medulloblastomas may show evidence of neoplastic astrocytes and they may show evidence of other kinds of neoplastic neuronal cells. If you look critically at descriptions of tumors in the nervous system in various locations you can find descriptions that fit all of those criteria which have been described for the cerebellar medulloblastoma.

So to my simplistic way of thinking, I see no reason why we should separate those in the cerebellum as some kind of special tumor and try to isolate the ones in other areas and give them other names. It does not seem logical to me. I understand that there is considerable resistance to this, and contrary to what may appear to be the case, I think the site of the tumor is extremely important. I certainly do not down-grade the importance of site of the tumor, and I'm sure that all of those features can be part of the diagnosis and have some kind of prognostic significance. However, today we are not addressing specifically the matter of prognosis.

The basic question is "What do we mean by the concept of primitive neuroectodermal tumor? How does this fit in with what we have observed?" There is an enormous body of literature which, if one looks at it critically, fits very well into the entire concept.

DR. EARLE: It seems to me that we are at least finding some areas where we have agreement and areas where we still have disagreement.

DR. GILLES: There are several points I would like to raise here. I think it is time for us to make decisions about some of these points. First of all, what is a diagnosis? For a pathologist there are two steps at least in arriving at a diagnosis. First, we inspect our slides for the histologic information that they provide. Second, as a separate intellectual step, we take into consideration clinical information, such as age, sex, site, and perhaps a number of other factors. These are two separate intellectual steps. We have to decide whether we as pathologists are going to describe the clinical evolution of a tumor, or what we see in the slides in conjunction with what we know about the location of the tumor. These are two different steps.

Dr. Earle, perhaps I can help us focus our discussion a bit if I could get you to tell us what the goals of our reclassification are today. We have heard three goals mentioned in passing. We have heard many people speak to the goal of classification so as to understand pathogenesis. We have heard one person, perhaps two, speak to classification based on therapeutic response. We have heard merely one person speak to the goal of providing prognostic information. I do not think I have heard anyone speak to an equally important goal, perhaps the most important of all, the classification for etiologic understanding. We do not understand the etiology of these tumors. I think before we continue trying to make a consensus, we need to know what is the consensus and what is the goal that this consensus is supposed to address.

DR. EARLE: My own opinion is that all of these goals fit into our common goal. Although it has not been voiced here, when we went to Geneva and got involved with the WHO, one of our goals was to treat the problem like the atom bomb. It was called "trying to produce a nonproliferation treaty." All of us had had about six or seven different classifications imposed upon us that we were supposed to teach to poor helpless students. But we did feel that if we could communicate on a unified level, it would save us all a lot of trouble. Also, what we agreed on was not that much different from Dr. Rubinstein's second series fascicle, which appeared shortly before the WHO Blue Book. We were rapidly approaching, I thought, a unified classification of brain tumors, while recognizing that we would have to add to and delete from it as time passed.

If anybody has another opinion, please voice it.

DR. WECHSLER: Let me make three statements. I think the great advantage of the WHO classification of 1979 was to agree on two goals: 1) to provide a good morphologic tumor description, and 2) to introduce the concept of tumor grading. I think we should continue along this line. As far as new research in Germany is concerned, we have our own type of classification. In the old days tumor grading was accepted only for gliomas but not for other tumors. Nowadays, we are rather confident that tumor grading, as suggested in the WHO classification, is of great importance with respect to many tumors. Only a few years ago, I was rather satisfied with our neurosurgeons that they at least accepted that there are so-called *benign minitumors* and so-called *malignant minitumors*, but intermediate grades have been neglected.

We know that there are considerable number of new achievements which will lead to elucidation of many tumors, at least in two grades. One of the goals which we should continue to pursue is to give tumor grading importance equal to that of making morphologic tumor diagnoses, particularly with respect to what we call differentiation and anaplasia.

As far as the details of the morphologically oriented WHO classification are concerned, there are certainly some revisions necessary. However, during this session we have only discussed embryonal tumors and primitive neuroectodermal tumors. This is a rather complicated field. I believe that we should just wait and see what happens later when we are talking about the glioma group. Certainly the glioblastoma multiforme should be introduced into the glioma group. There I agree with Dr. Barnard's suggestion.

From my point of view we should not be committed to being in a hurry. All the new techniques and approaches which have been presented today, especially those illustrated by Dr. Bigner, really need some more time for consideration. So as far as a basic consensus is concerned, shall we agree to continue promoting the WHO classifications with respect to morphologic tumor diagnosis and tumor grading or shall we discuss further the issues which you consider as fine tuning, or if you will, a more aggressive tuning of the present classification?

With respect to history, you mentioned the first edition of the Armed Forces Institute of Pathology (AFIP) fascicles. This was a wonderful tumor atlas describing peripheral and central nervous system tumors. If we compare the first and second editions, the differences were really dramatic, both from the morphologic point of view and from the conceptual basis of tumor grading. I would predict that while we are together to discuss the 1979 WHO classification, some tuning may be necessary, but the differences will not be as great as between the first and second editions of the AFIP fascicles on CNS tumors.

DR. SOBIN: All of the WHO groups very clearly have set out to describe entities that the pathologist can diagnose when he or she looks in the microscope. This has been and still is the goal. I tried to draw the distinction between conceptual terminology and descriptve terminology. We want to find terms that we can agree on, that when we look in the microscope and have the clinical information, we can come out with a term that the surgeon will understand and know what to do, the neuro-oncologist will understand and know what to do, the epidemiologist will know where to put them, and the tumor registry will find a slot for it.

Those who are working on more sophisticated types of classifications won't replace this, but will have a format to let this grow on. For example, the classification related to therapy. That should not replace this classification, that should rest on its own merit. For every tumor that has a morphologic term, there may be a further breakdown as to how susceptible it is to a particular type of chemotherapeutic intervention.

Now one should take the same approach with etiologic classification. The etiologic classification of tumors has never come about as in infectious disease because of the long period of tumor genesis, but that does not rule out looking at etiology, after you have a morphologic classification. I think that is what we are after.

Grading is a separate issue. For example, there is histologic grading, which is what the general pathologist does when he looks at a colonic carcinoma and says, "It is moderately or poorly differentiated." That is histologic grading. Then, there is prognostic grading, which takes into consideration many factors such as staging, the extent of the tumor, and perhaps certain genetic aspects. That is an entirely different matter.

I think what has been a problem at this session, perhaps not a problem but at least very interesting, is that we have been going from morphologic entities to concepts, to grading, to histologic grading, and prognostic grading, and moving in and out of these various axes of classification. But we must remember that the primary purpose of the first Blue Book, and of any revision, is to describe morphologic entities.

DR. KLEIHUES: Before I actually came to this meeting, I asked myself, how will it be possible to reconcile these different opinions? I think Dr. Rubinstein and Dr. Rorke have just made their points very clear. Dr. Rubinstein sees the divergency, and sees many different tumors, with different lines of differentiation depending on the site where the tumor originates. He sees this diversity, and he hesitates. As a matter of fact, he probably dislikes very much putting them all together under one heading. He says that with the methods available one sees different lines of differentiation in these tumors, why should one at this point put them all together in one basket?

Dr. Rorke, on the other side, says they have something in common. Although they may originate at different points in the central nervous system, they have a similar biologic behavior and often are morphologically indistinguishable. So she sees no reason to make a detailed description of each of these when they are basically the same.

These are really very opposite positions because one of them sees the differences, and the other sees what these tumors have in common. My proposal would be that we should accept basically the term *primitive neuroectodermal tumor*. I still believe that Dr. Rorke's concept is a very stimulating and a very valid one. Although I share with Dr. Rubinstein the view that many of these tumors can indeed be separated into different entities, in many cases, this is just not possible. I think anybody who does daily work on classification of these tumors is aware of medulloblastoma-like tumors which are not in the cerebellum.

I remember when I worked with Professor Zülch twelve years ago, he often called this medulloblastoma-like tumor in the cerebral hemisphere a *pineoblastoma*, or as he said earlier, *medulloblastoma of the pinealis*. This indicates very strongly that even at that time, we had the feeling that these tumors have something in common. We cannot really distinguish them morphologically.

My proposal would be, to introduce and accept the term *primitive neuroectodermal tumor* as a heading for a number of tumors, which we must then try increasingly to differentiate. This may not be our own kind of work in the future. I am convinced that geneticists or molecular biologists, will help us to make more of a diversification of the entities in this large basket. I believe that the basic concept of primitive neuroectodermal tumor should be introduced into the classification. However, the terms *medulloblastoma*, *pineoblastoma*, and all the others should not be abandoned. As a matter of fact, we will probably add to those some other entities which are morphologically similar, but which may have distinctive features that the geneticists and the molecular biologists will discover in the future.

So my desire would be, yes, let us accept that term, not as a waste basket, but as a superheading for those brain tumors with a medulloblastoma-like population of precursor cells which have the ability to differentiate in various directions - that is the point. If you have a tumor like the central neuroblastoma where all tumor cells show neuroblastic differentiation, then it is not a PNET since it is already on the way to a specific differentiation. The same is true with ependymoblastoma, where you may have a tumor with a medulloblastoma-like cell population as well as ependymal differentiation. That is a PNET in my view.

However, you may have, and Dr. Rubinstein may show later, a tumor which has ependymal differentiation and then we can call it an ependymoblastoma. But that is a rare case in my experience. More frequently we find that we have a primitive neuroectodermal tumor with ependymal differentiation, and I see no reason why we should not accept this term proposed by Dr. Rorke.

DR. HASSOUN: Since I am here, I have to insist on reporting my limited experience. I would like to come back to pineoblastomas. Pineoblastomas are not like medulloblastomas. They are quite different because they show various criteria of differentiation in the sense of pineocytic differentiation. I do not think that pineoblastoma belongs to the group of PNET since these tumors show some differentiation somewhere. Furthermore we cannot overlook the importance of electron microscopy.

Secondly, concerning an eventual differentiation toward neurons, my question is, "Do we know whether pineocytes can transform into neurons?" I insist on asking this question because I am not sure that these highly differentiated tumor cells are able to transform into another kind of highly differentiated cell. Is it possible that in this little gland, many cell types undergo neoplastic transformation? Since you see in the glioblastomas transformation of different cell types, blood cells, Schwann, and glial, of course, why not in the pineal body? I am not sure that pure gangliogliomas of the epiphysis, are to be classified in the group of true pinealomas. I am not sure at all since ganglion cells exist normally in this gland. Why evoke an eventual differentiation from the pinealocytes?

Pure gangliogliomas are found in other parts of the brain, so why not in the pineal body? I have observed two or three cases of pure gangliogliomas of the pineal body, and I think it is possible that these tumors developed from ganglion cells first. They were adjunct tumors.

I would also like to comment concerning the differentiation toward normal or in an astrocytic direction in otherwise classical pineocytomas. Why evoke two types of differentiation from pineocytes? I am not sure that we know how to deal with complex and multidirectional tumors. They appear to develop from pinealocytes, astrocytes, and neurons, and why not?

In my small experience with electron microscopy in ten cases, I did not find this intermediate form between pineocytes and neurons. I dealt with either pinealomas or with gangliogliomas. The transitional form, which would be necessary to prove this transformation, was not present in the material which I studied. So, concerning the pure gangliogliomas of the pineal body, I suggest that they be classified with other gangliogliomas of the brain, and that we recognize that these differentiations do exist. But from what? It is my opinion that they probably do not differentiate from pinealocytes.

DR. RUBINSTEIN: I think that the original ganglioglioma that we described in 1970 showed typical areas of large pineocytic rosettes, which I assume you will accept as evidence of pineocytomas.

DR. HASSOUN: Yes, I do.

DR. RUBINSTEIN: It was a pineocytoma, there was evidence of that. And with it, there were a lot of ganglion cells and all the differentiating neuroblasts. Electron microscopy was done on these. It is on the basis of these transitional cases, which perhaps you may wish to call *mixed* because of the existence of pineocytic differentiation in part of the tumors and ganglion cell production in another field nearby, that we assume that neuronal transformation can take place. It is based on that argument.

DR. HASSOUN: I think the discussion of this is obviously well taken, but my question was, "Are we sure that there is transformation of one cell type or another into neurons?" Are we dealing with mixed tumor including tumor ganglion cells and typical pineocytoma rosettes, which correspond in electron microscopy to absolutely typical pineocytes, with all this structural differentiation in the same direction?

My question, I'm sorry to insist, concerns the relationship between these two kinds of highly differentiated cells.

DR. BURGER: It appears Dr. Kleihues did not spend enough time with Professor Zülch to learn about germ cell tumors, and he did not spend enough time with me to understand PNETs. So let me try to straighten this out.

The idea of Dr. Rubinstein's embryonal scheme and Dr. Rorke's PNETs is irreconcilable for the reason that in her presentation the scheme was one of transformation of undifferentiated cells and that the other neoplasms then were expressions of that original transformation. If they differentiated along neural lines, one had a neuroblastoma, et cetera. The scheme, as Lucien Rubinstein would like to believe, and I think quite reasonably, is that cells anywhere along this differentiating pathway, as long as they are mitotically active, could be struck by an event, and become a neoplasm.

If the term *PNET* had been used in a very general sense like embryonal tumor, I think we could use that as a heading. However, Dr. Rorke has used it in a very specific sense. To use this in this classification seems to me to endorse the idea that all these tumors are really derived from a primitive cell, which is hit by whatever event happens. We ought to leave open the possibility that a neuroblast can be hit. Dr. Rorke, correct me if I am wrong, but it seems to me that that term *PNET* as you used it in your presentation would make that eventuality impossible. Can a neuroblast become neoplastic?

DR. RORKE: The point is that when we look at a tumor that seems to be composed of an undifferentiated cell, we do not know whether that cell is carrying a genotype that is going to make it develop into a neuron. I cannot tell at what stage in the development of the cell it is being transformed. All I can say is that it looks like it is undifferentiated. If I had some way of recognizing that it was destined to become a neuroblast, or a glial cell, then I could say something more specific about it. But since I cannot, the evidence is that all I can see with the light microscope and all the techniques we have to identify these specific kinds of cells, provides me with no indication that it has the potential for doing anything. All I can say is that it is an undifferentiated cell.

The other thing I can say is that if I see a tumor composed of that kind of cell, plus something else that I can identify as a neural or glial differentiation, then I simply report my observation. I have no preconceived idea as to what its differentiating potential might be because I do not know. All I can say is I have this population of cells, which are undifferentiated according to all the criteria that we now have at our disposal. Then, I have another population that shows these different features, and so I am simply describing the cellular composition of the tumor in an endeavor to classify it without any preconceived ideas at all.

DR. BURGER: Do you think the term *PNET* or the concept excludes the possibility that a neuroblast can be transformed?

DR. RORKE: No. Of course not.

DR. BURGER: I am asking only because the scheme as you presented it suggested that the original event was an undifferentiated, uncommitted cell.

DR. RORKE: No, when speaking about the light microscopic appearance, I said I do not know whether that cell has a genotype that is going to send it off in the direction of a neural cell. I am trying to be as honest as possible in terms of what I can observe.

DR. BURGER: I understand the microscopy part, but this concept seems to be one which assumes that it is an undifferentiated cell. That is where I would take issue. I think that is what makes it difficult to reconcile. If that is what is meant by the PNET scheme, which is drawn in your published article, I do not see how you can reconcile that with the concept that maybe there is a true neuroblastoma.

DR. RORKE: I do not really see that there is a problem.

DR. JELLINGER: I want to express a deep concern. We are not isolated in neuropathology. We are, I think, in a position very similar to that of hematopathologists confronted with lymphoma. We can argue that there is one stem cell, and you all have seen how rapidly and how drastically the classification of the malignant lymphomas has changed. Every second month or every second week, a new entity is reported which behaves biologically different. We should be cautious about going that same way. You know the deep differences between Luke's and Collins' classification, or working classification. I think we should avoid this.

The question is, "How will the WHO group deal, for example with the lymphomas? Do they include all of these entities? Do they want to correlate these different forms of classification, which in part originate from pure morphologic terms, as well as from immunologic terms and from biologic terms?" I think we should go the way of compromise, and I would agree with Paul Kleihues that *PNET* might be the heading, with several subgroups which can be changed, or there can be additional subgroups according to new biologic information.

DR. SOBIN: That is a very good example of what went wrong with the lymphomas. There were concepts that were brought up and the concepts did not rely on old terms, the concepts invented new terms, *centroblasts, T-cell lymphomas, etc.*, and I think that is why we are in a lot of trouble with them. For these reasons we are not going to approach a Blue Book on lymphomas. We are recommending that each group use the system that is most commonly used in their quarter of the world, and we are going to show the Kiel Classification of Lymphomas and the working formulation as two examples. But the lymphoma people fell into the trap of inventing new terms to satisfy new concepts. I hope that we can avoid that.

DR. KEPES: With your permission, I would like to address a question which I think is much less controversial but, nevertheless, important. I am directing my remarks to Dr. Sobin and to the rest of you as well. We have three days to consider this. The question is, "What should we put in this Blue Book?" Up to now, we were talking only about classification. That is part of it, the histologic classification. The next part is the definition of the classification. And we have 122 pictures.

My question is, "Are they sufficient?" I am not talking about new classification; I am posing the simple question: "If a pathologist looks into the microscope and sees a brain tumor which is not full of artifacts, a good quality specimen, should he be able to find an example of this in the Blue Book, or must he turn to another kind of publication?" Since I am involved in general pathology, I find that our colleagues in pulmonary pathlogy and gastrointestinal pathology can use the Blue Book to a much greater extent than we can because almost everything they can see in the microscope is in the WHO book.

I would say this is not true for the Blue Book on CNS tumors. Take a simple example, oligodendrogliomas. There are several variants, not with respect to classification but just by appearance. The spindle cell variant that Professor Zülch emphasized, the signet cell variant that Dr. Rubinstein called to our attention, those with palisading, and the one with cells that look like baby gemistocytes. All of these are kinds of tumors that we must recognize, but we won't find them in the Blue Book.

So my question is, "Are we going to expand our activities to include pictures of all the variants that are not presently included?" I submit that, just like the main difference in the outside appearance of part of the second edition of the Armed Forces Institute of Pathology fascicle of Dr. Rubinstein's work, and the Kernohan and Sayre book, before you even open it, you look at the size. It is about three times as thick, and I think this is one step we'll have to follow.

DR. SCHEITHAUER: It seems that we are being drawn into some sort of a recognition of the designation *PNET*, and as I see it, there are four possible places it could be put. One, it could serve as a super-heading if it were intended to be equivalent to embryonal tumors. I do not think that would be correct because there are plenty of embryonal tumors that are not confusing at all, for small cell neoplasia. Two, it could be used as a synonym for the small, round cell embryonal tumors. However, I do not think there is much to be gained by that, but in that case it might be used as a footnote. It could be designated, these are indeed synonymous with what has been termed in the literature *PNET*. Third, it could become interwoven into the body of the diagnoses themselves, the six or eight, or however many entities there are, with PNET in parentheses. I would hope we would not go that far. The fourth would be to reserve the term - I like the term *non-saponifiable residue* - for tumors that just defy classification but are nonetheless small round cell neoplasms. Even in this case, as I pointed out earlier, I think that using the term *small cell embryonal tumor*, not otherwise specified, is perfectly serviceable.

So, in summary, at any one of these four levels, I think we have preferable alternatives to introducing the term, short of a footnote regarding their being synonymous with the round cell embryonal tumors. I read these papers as well, looked at the charts and so forth, and came up with the same feeling that there was an implication for a mechanism of transfer with the PNETs. Whether truly intended or not, that is certainly the way the paper came across to me.

I think that legitimizes something that we are in absolutely no position to legitimize. That is pretty much what I have to say about it.

DR. RUBINSTEIN: Which one do you prefer?

DR. SCHEITHAUER: If we are going to do anything at all, I think that the thing to do would be to present the embryonal tumors as they exist in this spectrum, grouping those that appear as small cell neoplasms, close enough together to permit the placement of an asterisk and a footnote, or a small paragraph indicating that in the recent literature, perhaps with few references, these have come to be considered synonymous with this, but not to put so much emphasis on them as to bring the concept in all its possible implications.

DR. FIELDS: I regret that we do not have unlimited time. It is a pity because there are still issues that have not been resolved. However, I believe we are a lot closer to consensus than I would have thought possible earlier in this session. I would hope, as we consider later the issue of grading, that we will start to examine some of the questions regarding etiology and other issues that have been touched on only briefly in this first session.

From the outset, in planning this meeting, it was my understanding that the goal that we should try to acheive here is a classification which will be useful not only to those around this table, but also to persons elsewhere who may not have available to them the kind of sophisticated techniques which are found chiefly in academic departments.

PART TWO

MALIGNANT GLIAL TUMORS

Moderators:

W. K. ALFRED YUNG

and

JOEL B. KIRKPATRICK

CHAPTER VII

Can Cytogenetic and Molecular Genetic Analyses of Malignant Human Gliomas Be Used Yet To Supplement Conventional Classification Schemes?

DARELL D. BIGNER, B. Vogelstein and S. H. Bigner,
Department of Pathology and The Preuss Laboratory for Brain Tumor Research, Duke University Medical Center, Durham, North Carolina

With the advent of improved cell culture and cytogenetic technology, particularly the introduction of chromosome banding procedures in the last 20 years, there have been a number of studies of leukemias and lymphomas which have led to the incorporation of cytogenetic findings in clinical diagnosis. The most notable of these include the 15;17 translocation of acute promyelocytic leukemia, the 8;14, the 2;8, and 8;22 translocations of Burkitt's lymphoma and the 9;22 translocation of chronic granulocytic leukemia (SANDBERG et al. 1984). Although there are now a number of well described and accepted cytogenetic abnormalities in solid tumors, the incorporation of them into classification schemes and their use in diagnosis and establishing prognosis is less well established than in leukemias and lymphomas. Nevertheless, several findings such as losses or deletions of chromosomes 22, 11, or 13 in meningiomas, Wilm's tumor, and retinoblastoma respectively (ZANKL and ZANG 1972), and double minute chromosomes in advanced neuroblastoma (BRODEUR et al. 1984) have been well established.

We have recently completed and published a karyotypic analysis on a series of 54 malignant human gliomas (five anaplastic astrocytomas, 43 glioblastoma multiformes, three gliosarcomas, two giant cell glioblastomas, and one anaplastic mixed glioma) with G banding and C banding study of direct chromosome preparations of enzymatically dissociated gliomas and short-term cell cultures derived from the tumors (BIGNER, SH et al, 1984; BIGNER, SH et al. 1986; BIGNER, SH et al. 1988b). These findings compared favorably with earlier studies without chromosome banding published by MARK (1971) and with smaller studies with banding from other laboratories (AL SAADI and LATIMER 1980 and 1983; REY et al. 1987a,b, and YAMADA et al. 1980).

The essential findings in these series of 54 tumors are summarized in Tables I and II. The 54 tumors could be grouped into three groups according to stemline karyotype designated

(I), (II), and (III). Twelve of the tumors had normal stemlines or only lacked one sex chromosome. Eight of the tumors had near triploid to near tetraploid stemlines with four tumors showing losses of chromosome 22 and six containing double minute chromosomes (DMs). The largest group, group (III), had near diploid stemlines with 26 tumors showing gains of 7, 19 with losses of chromosome 10, 13 with structural abnormalities of 9 and 19 containing DMs. When numerically analyzed the 54 tumors had the findings summarized in Table II. Over 80% contained gains of chromosome 7 and more than 60% had loss of chromosome 10. Forty percent showed structural abnormalities of 9p, 40-50% had DMs and 50% of the group (II) tumors had losses of chromosome 22. The distribution, according to ploidy, showed that 1) almost two thirds of the tumors were near diploid; 2) approximately 22% were normal diploid, i.e. had normal diploid stemlines or showed only loss of a sex chromosome; and 3) 15% were triploid-tetraploid.

Table I. Stemline Karyotypes of 54 Malignant Human Gliomas (from BIGNER, SH et al. 1988b)

(I)	12 tumors (8 GBM, 4 AA) had normal stemlines or only lacked one sex chromosome.
(II)	8 tumors (8 GBM) had near-triploid > near-tetraploid stemlines. 4 tumors lacked 2 copies of chromosome No. 22; and 6 tumors contained double minutes chromosomes.
(III)	34 tumors (27 GBM, 2 GSGBM, 3 GS, 1 MG, 1 AA) had near-diploid stemlines. 26 tumors had gains of chromosome No. 7, 19 tumors had losses of chromosome No. 10, 13 tumors had structural abnormalities of chromosome No. 9 and 19 tumors contained double minute chromosomes.

GBM = glioblastoma multiforme, AA = anaplastic astrocytoma,
GSGBM = giant cell glioblastoma, GS = gliosarcoma, MG = mixed glioma

The cytogenetic abnormalities were analyzed among the histologic types of tumors present but the number of cases in the different groupings was not sufficient to detect statistically significant survival relationships among any of the groups or parameters analyzed. Despite the lack of clear-cut relationships between karyotypic abnormalities and survival in this initial study, it should be noted that there is a technical problem in the study of lower grade gliomas, particularly low grade astrocytomas. Because of the infiltrative nature of all astrocytic gliomas into normal brain and the ability of normal and reactive astrocytes, inflammatory cells and endothelial cells to grow in short-term culture, in the absence of cytogenetic abnormalities it is difficult to distinguish normal non-neoplastic cells from neoplastic cells which fail to exhibit karyotypic abnormalities. A much larger series of anaplastic astrocytomas and low-grade astrocytomas must be studied to determine if the trend toward a much higher percentage of cytogenetic abnormalities in glioblastoma multiforme is a valid finding. If so, then the cytogenetic abnormalities listed in Tables I and II would become a way in which to further distinguish anaplastic astrocytoma and lower grade astrocytic gliomas from the clinically most aggressive form, the glioblastoma multiforme.

Table II. Chromosomal Characteristics of Malignant Human Gliomas (data calculated from cases in BIGNER, SH et al. 1988b)

	# of cases with abnormality/ # of cases evaluated	Percent
Gains of Chromosome 7	26/32*	(80%)
Losses of Chromosome 10	19/32*	(60%)
Structural abnormalities of 9p	13/32*	(40%)
Presence of Double Minute chromosomes	25/42	(60%)
Losses of Chromosome 22	4/8	(50%) of Triploid-Tetraploid Tumors
Ploidy		
Near-diploid	34*/54	(63%)
Triploid-Tetraploid	8/54	(15%)
Normal Diploid or Sex Chromosome Loss only	12/54	(22%)

*In 2 of the 34 near-diploid tumors, the chromosomes were too short and poorly banded to allow detailed analysis.

The cytogenetic abnormalities described above, particularly those involving losses of chromosome 10 (CAVENEE W, personal communication, 1988), led JAMES et al. (1988) in Cavenee's laboratory to use restriction fragment length polymorphism (RFLP) analyses to examine in greater detail and with different methodology the loss of heterozygosity for chromosome 10 found by cytogenetics. Loss of heterozygosity occurred in 28 of 29 glioblastomas but in none of the lower grade astrocytomas and anaplastic astrocytomas. The specific gene that is the target of these chromosome 10 abnormalities has not been identified. Losses of heterozygosity for chromosome 17p were also observed in 8 of 24 astrocytic gliomas, and these changes involved astrocytomas of all grades (James et al. 1988b).

Other studies of human medulloblastoma and childhood astrocytic gliomas from two different laboratories have shown that the most frequent chromosome abnormality in these pediatric brain tumors is loss of one short arm of chromosome 17 with duplication of the long arm, that is isochromosome 17q (Bigner, SH et al. 1988a; Griffin et al. 1988). The RFLP analyses in adult gliomas and the cytogenetic abnormalities in the pediatric brain tumors would indicate the potential importance of loss of sequences on chromosome 17p in the development of brain tumors.

The cytogenetic and RFLP analyses demonstrating of the losses of chromosome 10, and the presence of double minutes strongly suggest a grade-related occurrence of these abnormalities in the most malignant and clinically aggressive form of brain tumor, the glioblastoma multiforme.

One of the reasons for performing cytogenetic analysis of tumors is to seek structural changes useful for selecting cases for more in-depth study with molecular genetic methodology (Bigner, SH et al. 1988c). The three most common cytogenetic manifestations of gene amplification are DMs, intrachromosomal,homogeneously-staining regions (HSRs) and abnormally-banded regions (ABRs). DMs have been frequently observed in malignant gliomas (Mark 1971; Bigner, SH et al. 1988b), and even in many of the early studies published in the 1960s. DMs usually signify the presence of gene amplification, but only recently have the amplified genes been identified. Gene amplification is defined as at least a fourfold increase in gene copy number due to selective multiplication of a small (less than 2000 Kb) region of a chromosome. It is demonstrated technically by hybridization of tumor DNA to cloned gene probes. Control hybridizations are performed with genes located on the same chromosomal arms, to exclude uneven DNA loading and to distinguish true gene amplification from increased copies of the entire chromosome (Kinzler et al. 1987; Wong et al. 1986; Wong et al. 1987). The original observation by Libermann et al. (1985) regarding amplification of the epidermal growth factor receptor gene (EGFR) in malignant human gliomas in four of ten cases has been studied now in great detail in our laboratories in a series of 149 cases as shown in Table III (Bigner, SH et al. 1988d; Bigner, SH et al. 1988e; Wong et al. 1987). Forty-four of the 149 cases showed gene amplification as defined above. The most frequently amplified gene was that coding for the EGFR.

TABLE III Gene Amplification in Human Gliomas (These 149 cases include those reported separately in BIGNER SH et al. 1988d and BIGNER SH et al, 1988e)

Tumor Type	Total	No Gene Amplification	Gene Amplification			
			EGFR	N-myc	gli	EGFR & N-myc
Astrocytoma	2	2	0	0	0	0
Anaplastic Astrocytoma	28	25	2	0	1	0
Glioblastoma	111	70	36	3	2	0
Oligodendroglioma	4	4	0	0	0	0
Anaplastic Mixed Glioma	3	3	0	0	0	0
Anaplastic Oligodendroglioma	1	0	0	0	0	1
Total	149	104	38	3	3	1

Moreover, not only is the gene amplified, but it is often rearranged and has specific regional deletions which result in an abnormally small gene product (HUMPHREY et al. 1988). Recently a clear-cut relationship to tumor grade has been shown with regard to amplification of the EGFR gene in which amplification was seen in 15 out of 57 patients with glioblastoma multiforme while none of 22 patients with anaplastic astrocytoma contained amplification of the EGFR gene (BIGNER, SH et al. 1988e). Amplification of N-myc and a gene called *GLI* was seen in three cases. The GLI gene was isolated from a permanent cultured cell line derived from a human glioblastoma with DMs (KINZLER et al. 1987). The sequence of the gene showed that it encodes a 118 kD protein with five zinc fingers suggesting that the gene product binds to DNA (KINZLER et al. 1988). The GLI gene is normally found on chromosome 12q13 (KINZLER et al. 1987).

There is too little molecular genetic data available from childhood brain tumors to make much of a statement of any kind at this point, although both DMs and amplification of the C-myc gene have been observed in two pediatric brain tumors including cell lines derived from a supratentorial glioblastoma and a cerebellar medulloblastoma (FRIEDMAN et al. 1988).

In summary, the cytogenetic abnormalities of malignant human gliomas, particularly glioblastoma multiforme, have now been well defined. However, there are still technical obstacles to incorporating cytogenetic data into classification schemes because of the difficulties in obtaining satisfactory preparations from lower grade tumors. It might be productive to launch a large prospective study to evaluate the diagnostic and classification significance of these cytogenetic abnormalities. Analysis of amplification of the four genes described above and RFLP studies of chromosomes 10 and 17 can be carried out using standard recombinant DNA techniques. Although the same technical problems with lower grade tumors also apply to these molecular techniques, *in situ* hybridization may provide a means to circumvent these problems as has already been shown to be applicable to the study of gene amplification in gliomas (WONG et al. 1987; BIGNER, SH et al. 1988d). Several of the cytogenetic and molecular genetic findings noted above are sufficiently well-established to justify carrying them forward into large scale multi-institutional trials, both to evaluate mechanisms of tumor evolution and to test their potential diagnostic significance.

CHAPTER VIII

Immunohistochemical Assessment of Differentiation and DNA Replication in Human Brain Tumors

P. KLEIHUES, A. Aguzzi, T. Shibata, O.D. Wiestler,
Laboratory of Neuropathology, Institute of Pathology, University of Zürich, Zürich, Switzerland

Introduction

The introduction of immunohistochemical methods has had a profound impact both on the science and practice of neuropathology. In many ways it has changed our way of thinking about neuro-oncogenesis (RUBINSTEIN 1987a), and in surgical pathology, it has led to a significantly higher level of diagnostic accuracy (BONNIN and RUBINSTEIN 1984; KLEIHUES et al. 1987). However, the application of immunocytochemical markers has not profoundly affected our view on the current classification of brain tumors (ZÜLCH 1979) although in a few cases it led to the abolishment of established entities (e.g., monstrocellular sarcoma) and the introduction of new diagnostic terms (e.g., the polymorphic xanthoastrocytoma). It should, however, be kept in mind that with respect to the histogenesis of neoplasms, the use of lineage-specific marker antigens may lead to erroneous interpretations. In contrast to their normal counterparts, neoplastic cells may undergo unpredictable changes in gene expression. This may result in the lack of a gene product typically present, or in the synthesis of a marker protein not usually expressed, in the cell of origin. Also, the panel of antibodies currently available offers no help in distinguishing normal from highly differentiated neoplastic cells. In this contribution we wish expand on two problems in which immunohistochemical markers have become increasingly useful, i.e. the assessment of cell kinetic parameters and analysis of the extent and direction of differentiation in embryonal CNS tumors.

Differentiation in the medulloblastoma

It has long been recognized that the cerebellar medulloblastoma, although usually appearing as an undifferentiated round cell tumor, may have the potential for neuronal and glial differentiation (RUBINSTEIN 1975). However, conventional histology does not permit quantitative estimation of the extent and frequency of this phenomenon.

This is particularly true for incipient differentiation which may manifest itself in the expression of lineage-specific marker proteins but still lacks unequivocal morphologic hallmarks. Our laboratory is privileged in having been chosen as the pathology reference center for an international medulloblastoma therapeutic trial, currently being conducted under the auspices of the (German) Society of Pediatric Oncology (GPO) and the Société Internationale d'Oncologie Pédiatrique (SIOP), which will eventually comprise approximately 300 cases (NEIDHARDT et al. 1987). In addition to conventional histologic evaluation, we have applied a variety of immunohistochemical markers with the objective of determining, more precisely than hitherto possible, the incidence of neuronal and/or glial differentiation in this important childhood embryonal tumor. Results obtained on the first 50 cases have previously been published (BURGER et al. 1987). We now report on a total of 246 evaluated surgical biopsies (Table I).

Table I Differentiation in the medulloblastoma

Neuronal differentiation		59.7 %
NSE positive clusters	50.4%	
Neuroblastic rosettes	42.1%	
Synaptophysin expression	39.6%	
Ganglioid cells	4.1%	
NFP expression	2.2%	
Astrocytic differentiation		
GFAP expression		12.9%
Neuronal plus astrocytic differentiation		7.0%
Ependymal differentiation		0.8%
Desmoplastic variant		12.2%

Based on 246 cases of the ongoing SIOP/GPO medulloblastoma therapy trial MED 84 (789)

Neuronal differentiation

Historically, neuronal differentiation has been most conspicuous, since neuroblastic (Homer Wright) rosettes can easily be detected in approximately 40% of medulloblastomas. In addition to clusters of rosettes, we found rhythmic palisading of tumor cells which in our view represents longitudinal sections through rosettes which may extend over considerable distances. Immunocytochemical studies revealed that neuron-specific enolase (NSE) also constitutes a valuable marker for incipient neuronal differentiation (GIORDANA et al. 1983; VELASCO et al. 1985; BURGER et al. 1987). In our series, we found sharply delineated islands of NSE-positive cells in approximately 50% of the cases. Often, the nuclei of these cells had a more vesicular appearence and cellularity was usually decreased. Medulloblastomas with somewhat more advanced neuronal differentiation showed areas with diffuse expression of synaptophysin in the neuropil. During brain development, the expression of this neuronal marker protein correlates with synaptogenesis and terminal differentiation. It appears that it is an integral calcium-binding membrane protein of presynaptic vesicles in CNS neurons and of similar vesicles in neuroendocrine cells (WIEDENMANN and FRANKE 1985; WIEDENMANN et al. 1986). In contrast to SCHWECHHEIMER et al. (1987) we only found focal immunoreactivity, usually in conjunction with additional signs of neuronal differentiation. Ganglioid but still immature neoplastic nerve cells were usually associated with synaptophysin expression whereas clearly identifiable ganglionic cells also expressed neurofilament protein (NFP), predominantly in cell processes but occasionally also in the perikaryon. This is in accordance with the findings of VELASCO et al. (1985) but at variance with those of ROESSMANN et al. (1983) and FRIEDMAN et al. (1985) who reported positive NFP staining in small undifferentiated cells whose neuroblastic or neuronal nature was not obvious from their appearance in conventionally stained histologic sections. In our experience, there is a continuous line of neuronal differentiation, the earliest characteristic being the focal expression of NSE and formation of neuroblastic rosettes, followed by synaptophysin expression and the appearance of ganglioid cells and, eventually, mature ganglionic cells and expression of NFP (BURGER et al. 1987; KLEIHUES et al. 1987).

Glial Differentiation

Glial differentiation in the medulloblastoma is much more difficult to define. Since specific and reliable markers have not been identified for neoplastic oligodendrocytes and ependymal cells, anti-GFAP is at present the only antibody that allows the estimation of glial differentiation in primitive neuroectodermal tumors. Most medulloblastomas contain GFAP-positive cells but the distinction of neoplastic versus reactive cells can be very difficult, particularly when these are sparsely dispersed within the neoplasm. Accordingly, the reported incidence of glial differentiation ranges from absent or very uncommon (COFFIN et al. 1983; SCHINDLER and GULLOTTA 1983) to 50% (ROESSMANN et al. 1983) and more (PALMER et al. 1981; PASQUIER et al. 1983). We have adapted the typing of

GFAP-positive cells in medulloblastomas as proposed by MANNOJI et al. (1981). These authors define neoplastic GFAP-positive cells (type 1) as morphologically identical to the ordinary medulloblastoma cell, i.e. with a hyperchromatic nucleus and a very scanty cytoplasm. Reactive entrapped astrocytes show a sparse chromatin and well-developed cytoplasmic processes. In addition, there is an intermediate type with fairly rich cytoplasm but short, plump cell processes and it is assumed that at least some of these cells are neoplastic. When we applied these stringent criteria to a large series of medulloblastomas (Table I), we found unequivocal evidence of GFAP expression by neoplastic cells (type 1) in only 13% of cases, and this figure agrees well with the findings of others (MANNOJI et al. 1981; SCHINDLER and GULLOTTA 1983; KUMANISHI et al. 1985). In the the desmoplastic medulloblastoma, immunoreactivity may be restricted to the clear, reticulin-free islands but GFAP expression as such is, in our experience, not restricted to this variant (HERPERS and BUDKA 1985). In contrast to neuronal differentiation, GFAP expression by medulloblastoma cells is usually restricted to a few foci of dispersed cells, with no evidence of a progressive differentiation into mature astrocytes. In several cases we observed sharply demarcated areas of low cellularity in which tumor cells showed a clear perinuclear halo. This has previously been interpreted as reflecting oligodendroglial differentiation. However, when we applied immunohistochemical markers, we found that in these areas there was a consistent expression of NSE and synaptophysin, indicating a tendency toward neuronal differentiation. We never observed oligodendroglia-like structures which did not express neuronal marker proteins. Ependymal differentiation with perivascular pseudorosettes was observed in only two biopsies.

Implications for histogenesis and the concept of primitive neuroectodermal tumors

The histogenesis of the cerebellar medulloblastoma has remained an enigma ever since the initial description by Bailey and Cushing more than 60 years ago (for review see RUBINSTEIN 1975). It is now generally agreed that this neoplasm is embryonal, i.e. derived from an immature precursor cell population. Some authors maintain that it is most likely the external granular layer of the cerebellum which predominantly (or exclusively) gives rise to neurons of the internal granular layer. Accordingly, one would expect that neoplasms derived from the external granular layer are either undifferentiated or display features of neuronal differentiation. Others have proposed the hypothesis that this tumor type originates from the subependymal matrix zone and is, therefore, not restricted to the cerebellum but occurs at all levels of the CNS. There is indeed evidence that medulloblastoma-like tumors occur in the cerebral hemispheres and it has been proposed (HART and EARLE 1973) that this group of embryonal neoplasms should be named *primitive neuroectodermal tumors (PNET)*. The conceptual basis for this nomenclature is the assumption that PNETs share a common progenitor population, i.e. the subependymal matrix cells, and that their neoplastic transformation at various levels of the CNS leads to tumors with similar morphology and biology (RORKE 1983). Analogous to non-transformed matrix cells, neoplasms derived therefrom would have the potential for neuronal, glial and ependymal differentiation. Consequently, the medulloblastoma has been merged into a nomenclatural pool designated *PNET*, together with various other pediatric brain tumors, including the pineoblastoma, the cerebral and cerebellar neuroblastoma and the ependymoblastoma (RORKE 1985). The results obtained in our series of 246 biopsies (Table I) clearly indicate that in the majority of cerebellar medulloblastomas (approximately 60%) there is a tendency for neuronal differentiation. Particularly impressive is the hierarchy of morphologic and immunohistochemical markers, starting with NSE-positive clusters and/or neuroblastic rosettes, followed by synaptophysin expression and, ultimately, the appearance of ganglioid cells and mature neoplastic ganglion cells expressing NFP. In one patient surviving 15 years after surgical resection and chemotherapy, autopsy revealed a highly differentiated gangliocytoma as the only remnant of a medulloblastoma which in the original biopsy only showed small areas with incipient neuronal differentiation. In contrast, glial differentiation seems to play a minor role. We observed GFAP-positive cells in about 13% of cases but only regionally in form of small clusters. The vast majority of tumor cells remained GFAP negative and there was no apparent progression into morphologically identifiable mature neoplastic astrocytes. In fact, the frequency of GFAP-positive medulloblastoma cells must be considered considerably lower than typically observed in oligodendrogliomas (HERPERS and BUDKA 1984; NAKAGAWA et al. 1986). On the other hand, morphologic features previously interpreted as reflecting oligodendroglial differentiation turned out to be immunoreactive for neuronal markers, and ependymal differentiation was negligible. In conclusion, our data would not be incompatible with the assumption that the medulloblastoma is indeed derived from the external granular layer of the cerebellum.

Cell proliferation in central nervous system neoplasms

In the histopathologic examination of surgical biopsies, the main emphasis is placed on an exact disease classification. However, in neoplastic diseases, additional information is required on the prospective biologic behavior of a lesion since this is known to vary considerably, even among neoplasms belonging to the same histologic type. The classical approach to this problem is the assessment of cellularity, nuclear pleomorphism, degree of differentiation, chromatin density and nucleolar size. In gliomas, the presence of vascular proliferation and necroses closely correlates with progressive anaplasia and poor prognosis. However, the most prominent feature of cells undergoing neoplastic transformation is a loss of growth control. Accordingly, pathologists have always shown a strong interest in the estimation of the rate of cell proliferation in biopsies from human neoplasms, and this is particularly true for slowly growing but ultimately recurring CNS lesions. Unfortunately, the mitotic index cannot be regarded as a reliable parameter since the mitotic phase accounts for less than 10% of the total cell cycle time, and during the interval between tissue removal and fixation, cells may complete mitosis without new cells entering the mitotic phase. ^{3}H-thymidine autoradiography offers a precise determination of the fraction of cells currently engaged in DNA synthesis (S-phase) but for ethical and economical reasons (preoperative injection of large amounts of ^{3}H-thymidine), application of this method is limited to a small number of selected patients (Hoshino and Wilson 1979) or in vitro incubations of excised tissue fragments (Bookwalter et al. 1986). More recently, immunohistochemical methods have become available which allow cell kinetic parameters to be routinely assessed in surgical biopsy specimens (Table II).

Comparison of immunocytochemical methods

Bromodeoxyuridine (BrdU) is a nucleoside analogue that is incorporated into nuclear DNA instead of thymidine and thus selectively labels cells in the S-phase of the cycle. It can be readily detected in ethanol-fixed, paraffin embedded tissue sections using a monoclonal antibody to BrdU (Gratzner 1982) in conjunction with classical immunoperoxidase histochemistry (Nagashima et al. 1985). Patients receive an infusion of BrdU prior to or during surgical intervention. Since BrdU has previously been used therapeutically as a radiosensitizer (Sano et al. 1968), its use is still licensed in most countries. At the dose required for cell kinetic studies (<200 mg/m^{2}), myelotoxic effects are unlikely to occur. Accordingly, this method has been increasingly used during the past five years, particularly in the laboratory of Hoshino and co-workers (Yoshii et al. 1986; Hoshino et al. 1986a; Hoshino et al. 1986b; Murovic et al. 1986; Nagashima et al. 1988). However, this technique has the inherent disadvantage

Table II Immunocytochemical methods for the identification of proliferating cells

	BdrU	Ki-67	Pol-α
Preoperative injection	BrdU[a]	-	-
Tissue fixation	Ethanol	Acetone[b]	PLP[c]
Tissue processing	Paraffin	Frozen/Squash	OCT[d]
Labeled cycle phase	S	G_1, S, G_2, M	G_1, S, G_2

[a] One-hour i.v. infusion of BUdR (150 to 200 mg/m^2) before or during the time of operation but before tumor biopsy (NAGASHIMA et al. 1985)

[b] 10 minutes at room temperature (BURGER et al. 1988)

[c] Fixation for 12 hr at 4°C in 2% PLP (2% paraformaldehyde, 10 mM NaIO4,, 75 mM lysine in 50 mM sodium phosphate buffer, pH 7.4). (For details see MUSHIKA et al. 1988)

[d] Embedding in OCT compound (MILES SCIENTIFIC, Naperville, IL, U.S.A); storage at -80 °C (MUSHIKA et al. 1988)

of requiring patient selection and patient consent. In addition, incorporation of BrdU may exclude molecular biologic studies on the same biopsy material by affecting the specificity of restriction endonucleases. As an alternative method, MORIMURA et al. (1989) have proposed to incubate small tissue fragments with BrdU *in vitro*, thus eliminating both patient selection and chemical DNA modification in the entire biopsy sample. However, *in vitro* penetration of BrdU into mammalian tissues is restricted (approximately 800 μm) and this limits the assessment of proliferating cells to a very small tumour volume.

The monoclonal antibody Ki-67 represents an attractive alternative to the BrdU method. This antibody recognizes an, as yet, unidentified nuclear antigen expressed in all continuously cycling cells, i.e. during G_1, S, G_2 and M phases of the cell cycle (GERDES et al. 1983; GERDES et al. 1984). Resting cells (G_0 phase) are consistently negative for Ki-67. Thus, in tumor biopsies the fraction of Ki-67 labeled cells directly reflects the size of the growth fraction. One would, therefore, expect that staining indices are considerably higher than those obtained with ^{3}H-thymidine autoradiography and BrdU, i.e., methods that selectively detect S-phase nuclei. However, comparison of the results with those obtained with ^{3}H-thymidine autoradiography (HOSHINO and WILSON 1979) and immunostaining for bromodeoxyuridine (HOSHINO et al. 1986a; HOSHINO et al. 1986b; YOSHII et al. 1986) fails to reveal significant differences. In our view, this is most likely due to different methods of quantitative evaluation. The data of Hoshino and co-workers appear to be based on evaluation at high magnification, thus favoring hot spots, whereas we have deliberately attempted to determine the Ki-67 index in large fields more likely to be representative for the entire neoplasm. In a combined *in vitro* study, MORIMURA et al. (1989), indeed, found 2-3 times higher labeling indices (L.I.) for Ki-67 than for BrdU ($L.I._{BrdU} = 0.34 \times L.I._{Ki\text{-}67} + 0.99$; r=0.92). Since immunoreactivity to Ki-67 is lost during formalin fixation and ethanol dehydration, application of the method is restricted to frozen sections and cytolologic preparations. An alternative methodology is the AMEX technique (SATO et al. 1986) which uses acetone as fixative and methylbenzoate/xylene for clearing, followed by paraffin embedding. In our experience, the Ki-67 immunostaining is particularly suitable for cytologic and squash preparations obtained from stereotactic brain tumor biopsies (OSTERTAG et al. 1987). It requires very small amounts of tissue and allows assessment of the growth fraction in several portions of the neoplasm without undue increase in workload. There is a close correlation between the Ki-67 labeling index in tissue sections and that determined in smear preparations from the same biopsy, the correlation coefficient amounting to 0.96 (BURGER et al. 1988). The major advantage over the BrdU method is the applicability to any surgical biopsy without the need for patient selection and pretreatment. Ki-67 immunostaining has, therefore, been used by several laboratories to estimate the proliferative potential of human brain tumors (BURGER et al. 1986; LANDOLT et al. 1987; ROGGENDORF et al 1987; GIANGASPERO et al. 1987; SHIBATA et al. 1988; ZUBER et al. 1988). More recently, MUSHIKA et al. (1988) introduced the use of a monoclonal antibody against DNA polymerase α. This enzyme is localized in the nucleus at the G_1-, S-, and G_2-phase (and in the cytoplasm during the mitotic phase) of cultured human cells, thus sharing a similar pattern of cell cycle expression with Ki-67. However, quantitative data comparing labeling indices obtained with both methods are not yet available. Immunoreactivity to anti-Pol-α is somewhat more resistant to fixation (Table II) but retrospective analyses of formalin-fixed routine biopsies are not possible.

Correlation with tumor classification and grading

Over the past few years, our laboratory has used Ki-67 immunohistochemistry in a variety of neurosurgical biopsies. The results obtained in 178 cases are shown in Table III. For each type of neoplasm, the mean Ki-67 index and the range

TABLE III Ki-67 expression in 225 neurosurgical biopsies

Tumor type	WHO Grade	Number of cases	Ki-67 staining index Mean	Range
Metastatic lesions		15	20.2%	(2.2-56.7)
Medulloblastoma	IV	3	13.8%	(10.2-19.4)
Glioblastoma multiforme	IV	65	9.9%	(1.2-27.6)
Astrocytoma, anaplastic	III	7	5.4%	(1.6-12.4)
Oligodendroglioma, anaplastic	III	8	6.1%	(1.7-16.9)
Oligo-Astrocytoma, anaplastic	III	3	13.0%	(10.9-16.8)
Chor. Plexus papilloma, anapl.	III	2	10.1%	(2.9-17.4)
Schwannoma, anaplastic	III	1	13.0%	
Meningioma[a]	III	5	4.6%	(2.9-8.7)
Astrocytoma	II	15	1.6%	(0.4-3.2)
Oligodendroglioma	II	14	2.6%	(0.5-9.5)
Oligo-Astrocytoma	II	6	1.9%	(1.3-2.9)
Ependymoma	I-II	5	0.9%	(0.2-1.6)
Astrocytoma, pilocytic	I	18	1.0%	(0.3-2.6)
Meningioma	I	18	1.4%	(0.1-3.8)
Schwannoma	I	6	1.2%	(0.3-2.7)
Pituitary adenoma	I	34	1.1%	(0.2-3.7)

[a] Hemangiopericytic, anaplastic and recurrent meningiomas

observed are reproduced. As expected, highest values were found in metastatic carcinomas although the Ki-67 index varied considerably in this group, ranging from 2.5 to 57%. This was similarly true for the most malignant type of human glioma, the glioblastoma multiforme. The mean staining index was below that of metastatic lesions but there was a broad variation ranging from 1 to 28%, with no evidence of a bimodal distribution. A large growth fraction was also found in malignant pediatric brain tumors, e.g. the medulloblastoma and malignant choroid plexus papilloma. Among astrocytomas and oligodendrogliomas, the Ki-67 index was found to parallel histologic grade as defined in the WHO classification (ZÜLCH 1979) and the known biologic behavior of the respective lesions. The correlation between Ki-67 index and WHO grading was particularly evident in the case of astrocytomas, with mean staining indices of 5.4% for the anaplastic variant, 1.6% for differentiated astrocytomas (fibrillary, protoplasmic and gemistocytic) and 1.0% for the juvenile pilocytic astrocytoma. Of the four ependymomas investigated, two were of the spinal myxopapillary type (WHO grade I) and had a very small growth fraction of only 0.2 and 1.1%. Although pituitary adenomas showed a low mean value (1.1%), there was considerable variation among the 34 cases investigated, the highest labeling index (3.7%) being found in an ACTH-producing adenoma associated with Cushing's disease (LANDOLT et al. 1987). In benign schwannomas, the staining index was generally low and labelled cells were usually found in tumor portions with Antoni A architecture rather than in the degenerative Antoni B areas. Similar results were reported from other laboratories (Giangaspero et al. 1987; Zuber et al. 1988). Most authors agree that despite the high degree of individual variation, the mean values for each tumor type correspond well with histologic classification and conventional tumor grading.

Prognostic value of cell cycle parameters

Although immunostaining for Ki-67 represents a convenient and rapid method for the estimation of the growth fraction in human central nervous system tumors, its future routine application would only be meaningful if the results obtained are of prognostic value and helpful in the planning of postoperative chemo- and radiotherapy. In meningiomas, a correlation between clinical behavior and Ki-67 index is emerging from our data and those of Roggendorf et al. (1987) as well as from the BrdU study reported by Hoshino et al. (1986b). In these studies, significantly higher labeling indices were found in the hemangiopericytic variant, in anaplastic meningiomas and in recurrent lesions. Since meningiomas usually constitute spherical lesions with insignificant cell loss, the BrdU labeling index can actually be used to calculate the approximate tumor doubling time (Cho et al. 1986). A less convincing correlation seems to exist in pituitary adenomas (Landolt et al. 1987; Landolt et al. 1988). These benign neoplasms had Ki-67 indices ranging from 0.2 to 3.7%. Invasive adenomas demonstrated significantly higher values (mean, 1.5%) than non-invasive adenomas (0.6%). However, gliomas represent a greater challenge for the prediction of biologic behavior and this is particularly true for the glioblastoma in which changes in the total tumor mass may depend as much on cell loss as on the proliferative potential of viable tumor cells. In an *in vitro* autoradiographic study using ^{3}H-thymidine, Bookwalter et al. (1986) found no correlation between labeling index and survival. This result is corroborated by a recent study from our hospital in which the Ki-67 index failed to be a prognostic factor for patient survival (Kleihues et al. 1989). A different picture may, however, emerge from more differentiated gliomas in which the assessment of the proliferative potential may aid in the decision on whether surgical resection should be followed by radio- and/or chemotherapy.

CHAPTER IX

Application of Immunohistochemistry for Tumor Grading in Human Neuro-Oncology

WOLFGANG WECHSLER and Guido Reifenberger, *Department of Neuropathology, University of Düsseldorf, Düsseldorf, Federal Republic of Germany*

Introduction

Histopathologic classification of tumors of the nervous system has to consider two major aspects: tumor diagnosis and tumor grading. Traditionally, both are evaluated by the neuropathologist using light microscopy and conventionally stained formalin-fixed and paraffin-embedded tissue sections. The currently employed WHO classification system for tumors of the nervous system has been based on this traditional approach (ZÜLCH 1979). Although the great majority of these tumors can be adequately diagnosed by means of conventional methods, the development of new immunohistochemical techniques together with a rapidly growing number of available specific polyclonal and especially monoclonal antibodies has provided a new powerful tool to improve diagnostic accuracy in histopathology. In contrast to the extensive work on differentiation antigens such as intermediate filaments and their usefulness for tumor diagnosis (KLEIHUES et al. 1987; PERENTES and RUBINSTEIN 1987; REIFENBERGER et al. 1987), comparatively few studies have been published which deal with the application of immunohistochemical methods as an aid to tumor grading. In the present report we will therefore discuss immunohistochemical approaches that might provide useful additional criteria for tumor grading.

Immunohistochemical detection of the MMA carbohydrate epitope

The monoclonal antibody MMA (anti-Leu-M1, Becton Dickinson Inc.), which was originally described as recognizing a differentiation antigen on human myelo-monocytic cells (HANJAN et al. 1982), recognizes the carbohydrate structure 3-fucosyl-N-acetyl-lactosamine (FAL) also known as X-hapten or stage-specific embryonal antigen I (SSEA I). This epitope shows a widespread distribution including certain hematopoetic cells, many epithelial cells and also a variety of cell elements derived from the neuroectoderm. There is evidence that FAL may play a role as a cell adhesion molecule for glial cells (NIEDIECK and LÖHLER 1987).

In the normal adult human brain we were able to detect MMA-immunoreactivity on a subset of astrocytes, oligodendrocytes, most ependymal cells and several neuronal cells located in distinct brain areas (MAI and REIFENBERGER 1988). In addition, we found that reactive astrocytes located in the periphery of brain tumors and within the peritumorous edematous zones exhibited an increased amount of immunoreactivity. Investigating a series of various human brain tumors by means of immunohistochemistry on paraffin sections we found that the majority of astrocytomas, oligodendrogliomas, mixed gliomas, glioblastomas and medulloblastomas were MMA-negative (SZYMAS et al. 1987b). Only pilocytic astrocytomas and individual WHO grade II astrocytomas, some oligodendrogliomas, mixed gliomas and ependymomas were MMA-positive. Other tumors investigated, such as choroid plexus papillomas, neurinomas, meningiomas, and craniopharyngiomas were all MMA-negative. Thus immunohistochemistry for MMA appears to be of considerable value not only for the distinction between different tumor types but also for discrimination between reactive and neoplastic brain tissue since reactive and persisting glial tissue always stained strongly, whereas, at least in the group of malignant gliomas, tumor cells remained constantly negative. In addition, it could provide the neuropathologist with some information on the biologic behavior of gliomas because all high-grade gliomas (WHO grade III and IV) were consistently MMA-negative, while MMA-positive gliomas all belonged to the low-grade group (WHO-grade I and II).

Immunohistochemical Detection of Protein Kinase C

Protein kinase C (PKC), the major receptor for phorbol ester tumor promotors, is a phospholipid- and calcium-dependent serine/threonine-specific phosphorylating enzyme which plays a major role as an intracellular mediator of fundamental processes such as growth, differentiation, activation of cells by hormones, growth factors and neurotransmitters, and regulation of gene expression (NISHIZUKA 1986). In a recent study we investigated 92 human brain tumors by means of immunohistochemistry on frozen sections using the mouse monoclonal IgG-2a antibody C5 recently characterized by YOUNG et al. (1988) and commercially available from Amersham-Buchler, Braunschweig, FRG. This antibody recognizes an epitope on the α and β forms of bovine protein kinase C. In normal adult human brain we found C5-immunoreactivity predominantly located in the cytoplasm and processes of astroglial cells, whereas oligodendrocytes, ependymal cells, choroid plexus cells, neurons, leptomeningeal cells and blood vessel walls always appeared negative. C5-immunoreactivity was also strongly present in reactive astrocytes and in a variable proportion of tumor cells in astrocytomas, oligodendrogliomas, mixed gliomas, ependymomas and glioblastomas. PKC-positive tunor cells were demonstrable in tumors of all WHO-grades but in double-labeling experiments with the proliferation-associated Ki-67-antigen (see below) we were able to demonstrate that most of the Ki-67-positive glioma cells were negative for C5. Thus, as a consequence, the average fraction of C5-positive glioma cells appeared lower in the group of high-grade gliomas compared with the low-grade ones. Since C5-immunoreactivity was absent in most non-glial neoplasms, immunohistochemrstry using this monoclonal antibody could be helpful for diagnostic purposes such as the distinction between glial and non-glial neoplasms, but could also be of some value for the grading of gliomas (REIFENBERGER et al. 1988b).

Immunohistochemical detection of epidermal growth factor receptor

Biochemical studies have demonstrated epidermal growth factor receptor (EGFR) expression in human anaplastic gliomas with evidence of EGFR gene amplification and rearrangement (LIBERMANN et al. 1984; 1985; FILMUS et al. 1985; HUMPHREY et al. 1987, WONG et al. 1987). Very recently, BIGNER, SH et al. (1988e) investigated a series of malignant gliomas for the expression of EGFR by means of *in situ* hybridization and immunohistochemistry. They found EGFR immunoreactivity in 14/15 malignant gliomas with, and in 11/16 malignant gliomas without, EGFR gene amplification. In contrast, EGFR mRNA was detectable only in the 15 tumors with gene amplification. We have recently investigated a series of 100 human tumors of the nervous system for EGFR expression by means of immunohistochemistry on frozen sections using a specific monoclonal antibody EGFR-1 (produced and characterized by WATERFIELD et al. 1982; commercially available from Amersham-Buchler, Braunschweig, FRG). EGFR-like immunoreactivity was strongly present in 23 out of 29 (79%) high-grade gliomas, i.e. anaplastic astrocytomas, oligodendrogliomas, mixed gliomas of WHO grade III and glioblastomas of WHO grade IV. In contrast, most low-grade gliomas investigated, including astrocytomas, oligodendrogliomas, mixed gliomas and ependymomas, proved to be negative. Only 2 WHO grade II astrocytomas out of 22 low-grade gliomas (9%) contained some EGFR-positive tumor cells. Despite this obvious relation of EGFR expression to WHO grade there was no close correlation to the tumor growth fractions as determined by Ki-67 labeling (see below). In the group of high-grade gliomas we observed individual cases of highly proliferative malignant gliomas which proved immunohistochemically to be EGFR-negative as well as other tumors which stained strongly for EGFR but had low Ki-67 indices. In addition to the gliomas, EGFR was detectable in smaller amounts in 13 out of 15 meningiomas and in varying intensity in intracerebral metastatic carcinomas. Other tumors, including 7 of 8 medulloblastomas, 2 cerebral primitive neuroectodermal tumors in children, 1 intracranial malignant melanoma, 1 malignant lymphoma, 1 ganglioneuroma and 3 benign neurinomas were EGFR-negative. In contrast, 1 malignant neurinoma (WHO grade III) revealed strong generalized immunoreactivity.

In summary, our data indicate that EGFR expression in human brain tumors depends on the tumor type, and in the glioma, and probably also the neurinoma group, is associated with high grade of malignancy. Besides their implications for modern concepts of tumor growth promotion these findings could be of considerable relevance with regard to tumor grading of human gliomas.

Immunohistochemical detection of tumor cells in S-phase by the BrdU-method

Gratzner (1982) developed a monoclonal antibody with specificity for bromodeoxyuridine (BrdU), a thymidine analogue which, like ^{3}H-thymidine, can be incorporated into DNA during the S-phase. BrdU is administered to the patient intravenously before the operation and afterwards detected on paraffin-embedded tissue sections using the monoclonal anti-BrdU antibody and immunoperoxidase or immunofluorescence methods. Compared with ^{3}H-thymidine, the BrdU method has the major advantage of being neither radioactive nor myelotoxic at least at the doses used for *in vivo* labeling (Russo et al. 1984). Furthermore, BrdU has been shown to act as a radiosensitizing agent, thus improving the effect of radiotherapy (Sano et al. 1968; Hoshino and Sano 1969). A large number of different human brain tumors have been investigated by the BrdU-method. Considerable data exists concerning the labeling indices obtained for individual brain tumor types (see Danova et al. 1988; Hoshino et al. 1985; Hoshino et al. 1986a; 1986b; 1986c; Hoshino 1987; Nakashima et al. 1985, Nagashima et al. 1986; Nagashima et al. 1988; Fukui et al. 1986). The extensive work by Hoshino and collaborators especially must be mentioned in this context. According to these authors the determination of the proliferative potential of human brain tumors by the BrdU method represents a valuable supplementary histopathologic method which allows a more accurate estimation of tumor grading. Unfortunately, most of the cited publications did not correlate the results obtained by the BrdU method with the WHO grading system which clearly complicates comparative analysis of these data with those obtained by other methods such as immunohistochemistry for the Ki-67 antigen.

Immunohistochemical detection of the proliferation-associated nuclear antigen Ki-67

Recently GERDES et al. (1983) developed the monoclonal antibody Ki-67 by immunization of Balb/c mice with a crude nuclear fraction of L428 cells. This antibody apparently recognizes a nuclear antigen, of yet unknown nature and function, present in the nuclei of human cells during the G_1-, S-, G_2-, and M-phases of the cell cycle but absent in G_0-cells (GERDES et al. 1984). Immunohistochemical demonstration of the Ki-67 antigen has therefore been applied to identify the proliferating cells in frozen sections of various human tissues and tumors, e.g. gliomas (BURGER et al. 1986; GIANGASPERO et al. 1987), meningiomas (ROGGENDORF et al. 1987; ROBSON et al. 1987) and pituitary adenomas (LANDOLT et al. 1987). In our own series of now more than one hundred brain tumors (DECKERT et al. 1988) (see table I) we found, in partial agreement with the studies mentioned above, a good overall correlation between the average values of Ki-67 labeling and the WHO grades of malignancy. Low-grade astrocytomas, oligodendrogliomas, mixed gliomas and ependymomas showed consistently lower labeling indices than anaplastic gliomas of WHO grade III and IV. The highest values for Ki-67 in our series were seen in metastatic carcinomas, medulloblastomas, and two outstanding cases of recurrent astrocytoma. There were, however, wide ranges in the staining indices especially in the glioblastoma, medulloblastoma and metastatic carcinoma groups. Interestingly, the average labeling indices obtained in anaplastic astrocytomas, oligodendrogliomas, and mixed gliomas of WHO grade III were higher than that obtained for WHO grade IV (the glioblastoma group). We interpret this finding as a consequence of the pronounced regional heterogeneity in glioblastoma multiforme tumors which might generate considerable sampling problems resulting in the wide range of Ki-67 indices in this tumor group. Another interesting phenomenon is noted in individual low-grade gliomas presenting with fairly high Ki-67 labeling indices. However, whether these cases will indeed demonstrate poorer prognoses remains to be elucidated in further follow-up studies.

In summary, the Ki-67 method, which requires frozen material, represents in our experience a useful aid for determination of the growth fraction in human brain neoplasms at the time of operation and should therefore become routine as a source of helpful additional information for the grading of these tumors.

TABLE I Ki-67 labeling indices in human brain tumors

Diagnosis	WHO-grade	No.	Ki-67 labeling indices		
			x (%)	STD	range (%)
Astrocytoma, pilocytic	I	4	<1	<1	0 - <1
Astrocytoma	II	13	1.2	2.1	0 - 7.9
Rec. Astrocytoma	II	3	1.7	2.4	0 - 5.0
Astrocytoma, anapl.	III	4	9.9	7.5	2.0 - 21.2
Rec. Astrocytoma, anapl.	III	1	10.1	-	-
Rec. Astrocytoma, anapl.	III/IV	2	48.5	4.5	44.0 - 53.0
Oligodendroglioma	II	3	2.8	0.3	2.5 - 3.2
Oligodendroglioma, anapl.	III	1	11.0	-	-
Rec. Oligodendroglioma, anapl.	III	6	14.4	3.6	10.4 - 19.5
Mixed Glioma	II	1	3.3	-	-
Mixed Glioma, anapl.	III	7	14.3	8.6	3.7 - 29.0
Ependymoma	II	2	2.0	1.2	0.8 - 3.2
Glioblastoma	IV	18	6.5	6.3	0 - 22.1
Rec. Glioblastoma	IV	5	8.1	6.9	0 - 17.6
Medulloblastoma	IV	4	19.9	13.1	6.0 - 36.4
PNET	(IV)	2	13.7	0.3	13.4 - 14.0
Neurinoma	I	3	0	-	-
Neurinoma, anapl.	III	1	22.0	-	-
Ganglioneuroma	I	1	0	-	-
Capillary Hemangioblastoma	I	1	<1	-	-
Meningioma	I	14	1.0	1.1	0 - 3.8
Meningioma	II	1	5.0	-	-
Rec. Meningioma	II/III	1	10.0	-	-
Metastatic Carcinoma	-	12	14.9	15.7	0 - 46.0

x (%) = mean value in percent; STD = standard deviation; Rec. = recurrent; PNET = primitive neuroectodermal tumor of the cerebrum.

Conclusion

Immunohistochemistry has become an established method alongside conventional light microscopy for the histopathologic differential diagnosis of tumors of the nervous system. In the present discussion we have focused attention on the possible applicability of immunohistochemistry for the grading of these neoplasms. Tumor grading in the past was based solely on traditional morphologic criteria for anaplasia such as high cellularity, poor cellular differentiation, nuclear hyperchromatism, elevated mitotic index, presence of atypical mitotic figures, pleomorphism of tumor cells and tissue architecture, necroses and abnormal vascular proliferations. However, none of these criteria can be regarded as an absolutely reliable indication of anaplasia or rapid tumor growth. Therefore their evaluation can frequently lead only to an approximative estimation of a tumor's grade of malignancy. Thus, new methods providing additional objective criteria are needed to improve the accuracy of tumor grading in neuro-oncology. First steps towards quantifiable criteria for tumor grading are provided by some of the immunohistochemical approaches discussed above. Supplementing and refining the traditional WHO system of four grades of malignancy should lead to a better estimation of the grade of malignancy of a tumor, which from the point of view of both the patient and the clinician is undoubtedly of great relevance. Finally, we would like to emphazise that the application of some of the methods discussed and also further techniques such as DNA-cytometry requires the availability of unfixed material. It should therefore become routine procedure to store a sample of each tumor as unfixed frozen material.

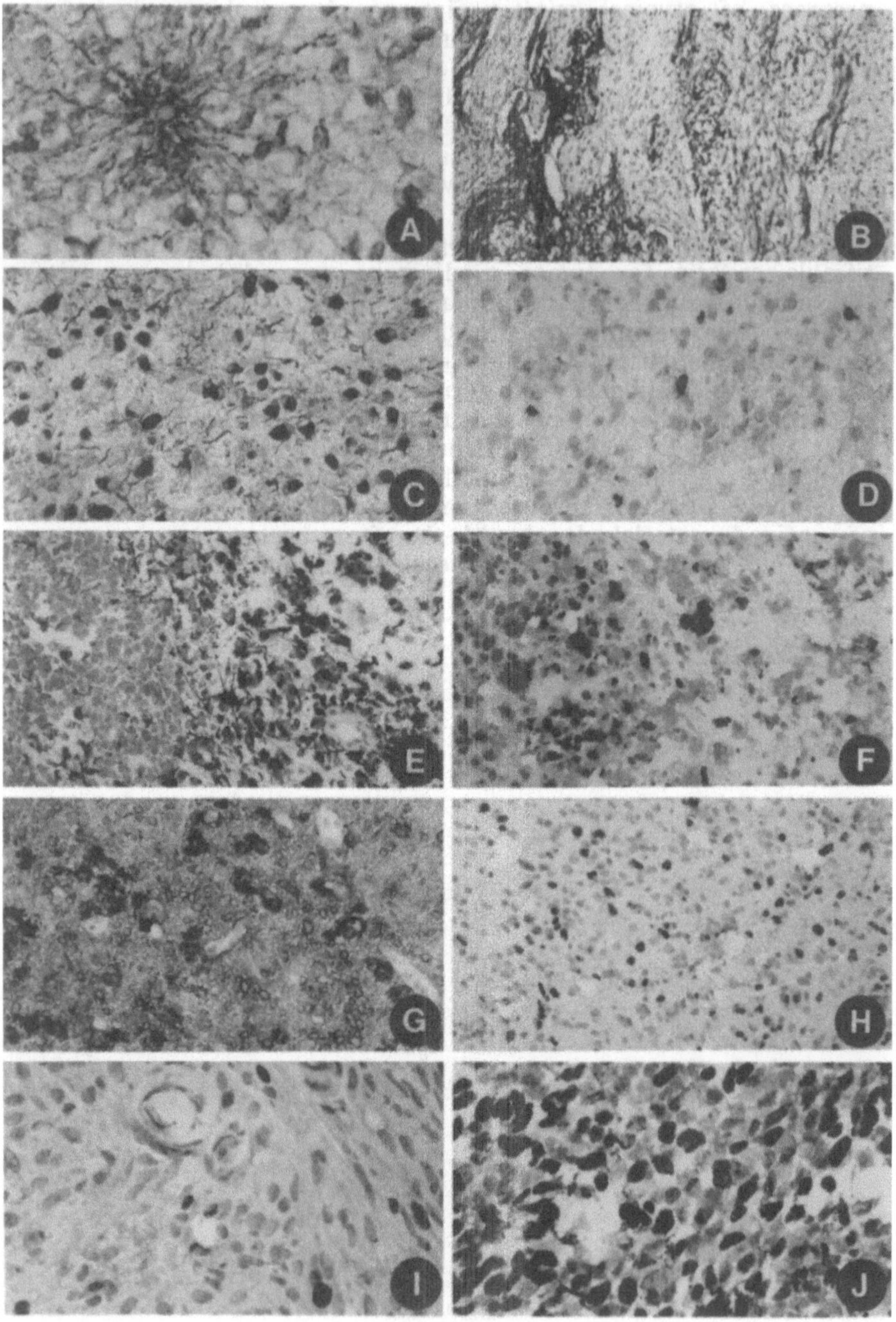
A
B
C
D
E
F
G
H
I
J

Figure 1:

A) **Oligodendroglioma (WHO grade II):** Strong staining of a reactive astrocyte for MMA. Adjacent tumor cells are MMA-negative (Np 220/84). 400 x.

B) **Glioblastoma (WHO grade IV):** Immunoreactivity for MMA is restricted to reactive and persisting glial tissue whereas glioblastoma cells are negative (NP 243/85). 60 x.

C,D) **Recurrent astrocytoma, gemistocytic (WHO grade II):** Recurrence five years after the first operation demonstrating the typical histologic appearance of a gemistocytic astrocytoma. C) Strong intracytoplasmic PKC-like immunoreactivity in numerous gemistocytic tumor cells. D) Moderate fraction of Ki-67-positive tumor cells. The mean labeling index of this tumor was 5% (NP 661/87). 150 x.

E,F) **Recurrent astrocytoma, highly anaplastic (WHO grade IV):** Recurrence 2.5 years after the first operation of an isomorphic gemistocytic astrocytoma of WHO grade II demonstrating a dramatic increase in anaplasia and therefore graded as WHO grade IV. E) PKC-like immunoreactivity is present in tumor cells of the more differentiated astrocytic part of the tumor on the right side whereas the densely packed small anaplastic glioma cells on the left are mostly negative. F) In the same tumor staining for Ki-67 reveals the opposite pattern showing numerous small anaplastic cells on the left intensely labeled whereas the number of positive cells on the right is lower. The mean Ki-67 labeling index of this tumor was 44% (NP 732/88). 150 x.

G,H) **Glioblastoma (WHO grade IV):** G) Immunoreactivity for EGFR is diffusely present on nearly all tumor cells while some tumor cells demonstrated increased expression. Note completely unstained blood vessels. H) The same tumor stained for Ki-67 showing numerous immunoreactive glioma cells. The mean labeling index of this tumor was 22% (NP741/87). 150 x.

I) **Meningioma, endotheliomatous (WHO grade I):** Tumor area with several tumor cells positive for Ki-67. The mean labeling index, however, was only 1% (NP725/87). 400 x.

J) **Medulloblastoma, desmoplastic (WHO grade IV):** Many tumor cells react positively for Ki-67. Mean labeling index was 36% (NP 256/88).400 x.

Discussion of Chapters VII-IX

DR. RUBINSTEIN: The three preceding papers are important and exciting. They deserve to be discussed before their impact is diluted by the next topic. Dr. Bigner has described his work on the analysis of restriction-fragment length polymorphism in gliomas in collaboration with Dr. Cavenee. We now have for the first time the opportunity to examine the role of suppressor, or regulatory genes in gliomas. This is an opportunity to look at cases of familial glioma to see whether there exists a constitutional expression of a karyotypic abnormality in the polymorphonuclear leucocytes of the relatives of these patients. Such an investigation is possible because there is a Registry of Familial Gliomas (TIJSSEN et al. 1982), which was made operative a few years ago in Leyden, The Netherlands.

As regards the paper by Dr. Kleihues, we entirely agree with his findings. Dr. Christos Katsetos, in our laboratory, has independently confirmed all his findings with synaptophysin. Dr. Katsetos has also used other neuronal markers, such as tau, Map 2, and an isotype of beta-tubulin, all of which are relatively specific. We can only support the findings of Dr. Kleihues that medulloblastomas show much more neuronal differentiation with these markers than could have been evaluated by the traditional staining methods. We also agree that the oligodendroglial-like areas in medulloblastomas seem to demonstrate evidence of neuronal differentiation with these markers. This can be shown not only with synaptophysin, but also with monoclonal antibodies against tau, Map 2, and an isotype of beta-tubulin. There are also undoubtedly tumor cells that are GFA protein-positive. We have to accept, at least provisionally, that the immunoreaction indicates astrocytic differentiation. Dr. Kleihues is right in being cautious on this point, but the expression, by tumor cells, of GFAP has been well documented. Moreover, it can occasionally be shown in medulloblastoma cells which have spread into the subarachnoid space, or even extraneurally. In those sites it cannot be argued that one is dealing with preexisting reactive astrocytes. In summary, then, we have to accept the finding that bi-directional differentiation of medulloblastoma may occur and that GFAP remains what has been called hard evidence of astrocytic differentiation.

I was extremely intrigued by the demonstration of S-100 protein in the dysplastic glial cells of central neurofibromatosis. The explanation suggested by Dr. Kleihues, i.e., to relate this finding to an arrested development of these cells, seems to me most reasonable.

Finally, the demonstration of the proliferative activity of tumor cells using Ki-67 or a monoclonal antibody against BrdU is one of the most promising areas in brain tumor biology today, and clearly indicates the path to follow.

These were three excellent papers and we are greatly indebted to Drs. Bigner, Kleihues, and Wechsler.

DR. WECHSLER: I would like to comment on Dr. Kleihues' presentation concerning the medulloblastomas. I think it was mentioned that approximately 50% of the tumor cells represent some kind of neuronal differentiation as exemplified by an immunoreactivity for NSE, synaptophysin and neurofilaments. What about the nature of the rest of the tumor cells, which you call undifferentiated? Do you see any markers available for this cell population? Have you, for instance, tried the pan-neuroectodermal marker used by Dr. Bigner in his report on medulloblastoma cell lines?

DR. KLEIHUES: The study that we carried out was based on slides and paraffin blocks that were sent to us from all over Europe. We had no influence on the mode of fixation and we were unable to obtain fresh biopsy material. We therefore had to rely on markers known to yield reliable results on formalin-fixed, paraffin-embedded tissue. Unfortunately, monoclonal antibodies against fixation-sensitive antigens could not be used, e.g. the proliferation marker Ki-67. With the immunohistochemical methods available, we found no evidence for neuronal or glial differentiation in approximately 25% of medulloblastomas.

DR. WECHSLER: In our study on medulloblastomas, the undifferentiated cell population demonstrated, at least in some of our cases, a relatively widespread cytoplasmic immunoreactivity for Leu 7 (REIFENBERGER et al. 1987)

DR. DAVIS: Paul Kleihues' results, of course, are very attractive to me because they agree with my bias. I was at the International Association of Pathologists meeting in Washington a few weeks ago, where there were two reported studies of synaptophysin in medulloblastomas. One group reported almost 100% positivity, and the other a smaller percent, which I do not recall at the moment. The other group from Minnesota had about 30% positivity with GFAP in their medulloblastomas, so I think we just need to be a little bit cautious at this moment about real numbers.

DR. KLEIHUES: We observed focal expression of synaptophysin in up to 40% of the cases. The very divergent data on GFAP expression in medulloblastomas are probably based on the fact that different criteria have been used to identify GFAP-positive neoplastic cells. It now appears that those authors who used the stringent criteria first defined by MANNOJI and al. (1981). observe focal GFAP expression in approximately 10-15% of cases.

With respect to synaptophysin, there are some conflicting results. SCHWECHHEIMER et al. (1987) reported 100% expression and state that it is a reliable marker. However, one needs resin blocks. We do this in a rather stringent way and we see clear foci of synaptophysin present in only 20-25% of the cells.

DR. RORKE: I would like to comment on some of the chromosomal studies. We are fortunate to have at the Children's Hospital in Philadelphia a cytogeneticist who was trained in Dr. Bigner's laboratory and we are attempting to do chromosomal studies on all childhood brain tumors that are operated on in the hospital. To date, we have studied about 20 medulloblastomas and have also found that the most consistent chromosomal defect seems to be the iso 17Q. There is also a smaller cluster of tumors with abnormalities on chromosome 6.

We had one remarkable example that showed some very intriguing findings, particularly in view of Dr. Sandra Bigner's studies on the double minutes in malignant gliomas. A posterior fossa tumor in a child showed a pattern that was very similar to one that Professor Kleihues showed in which there were large areas of undifferentiated cells, and a separate population of neoplastic cells that looked different.

These appeared to be swaths of neoplastic astrocytes forming rivers through the bands of differentiated cells. The chromosomal studies on that tumor were fascinating because there were many double minutes. So I think that increasing numbers of chromosomal analyses of the tumors show promise of giving us very exciting information relative to what the genotypic abnormalities are.

I would like to make another comment relative to Professor Kleihues' presentation. I thought it was very beautiful, and I would say that over the past few years, I have been coming to the same conclusion: i.e., that what have been interpreted as islands of oligodendroglioma, really are not, but more likely represent neural differentiation.

DR.KLEIHUES: I would like to address a question to Dr. Bigner regarding his studies on the genetic basis of malignant transformation of glial cells, depicting over-expression of genes or deletion of genes. Have you also looked for dominant astrocytes, genes, or whatever?

DR. BIGNER: We have not done a study such as Mark Israel is presently doing. He showed some preliminary results, at the recent Preuss Foundation meeting, of a screen of *expression* with many gene probes, i.e., switched on or off genes in glial tumor cell lines. The problem that we get into is interpretation, and our molecular biologist, Bert Vogelstein, always says "expression relative to what?" Unless one has a control cell population and can define expression in malignant cells in a quantitative sense relative to that in control cells, one has a great deal of difficulty in determining simply *expression* of an oncogene.

Now what Mark is turning up, as he has screened some 40 or more glioma lines, are some interesting patterns of simply switched on or off oncogenes. I am sure within a year or so we will see the published results of that work. What we did was to seek amplification, with all the known 20 or more oncogene probes, focusing specifically on tumors and cell lines with cytogenetic hallmarks of amplification, the presence of double minute chromosomes. There are still three or four cases in which double minute chromosomes, implying gene amplification, are present that we think are amplifying some gene that is not any of the known oncogenes. These cases will have to be analyzed in a laborious manner as was done with the Roninson assay to identify Gli.

DR. KLEIHUES: With modern molecular biologic techniques, amplified and overexpresssed oncogenes can readily be identified. However, the activation of proto-oncogenes may be due to more subtle changes such as point mutations. In such cases you would probably have to rely on transfection assays?

DR. BIGNER: This is only the "tip of the iceberg" of what has to be done. As you know well, the difficult thing, and what may turn out to be important in all this, as the experimental neu oncogene story is indicating, are point mutations. Point mutations with overexpression are very real phenomena, but finding them is like searching for "a needle in a haystack" until you have identified a place on which to focus. My opinion is that glioblastoma multiforme is a final common pathway and that there will be different mechanisms of transformation and possibly different etiologies. We are going to have to laboriously sort out these potentially different transformation mechanisms. I would not be at all surprised if a significant mechanism turns out to be point mutation of specific gene families. However, there is no evidence at this point that such point mutations exist with any significant frequency in human gliomas.

DR. YUNG: There was a recent paper which reported the screening of TGF alpha and EGF beta genes in cell lines from a large number of tumor specimens (DERYNCK et al. 1987). What the authors found was that in only one glioblastoma cell line that they screened was there any amplification of TGF alpha. In my laboratory we have done a small screening study of some glioma specimens and we have found the same thing. The GFA receptor is the only one to amplify in some of these tumors. Surprisingly, we found two tumors that had amplified TGF alpha and both were recurrent tumors. We did not find expression of TGF alpha in any of the untreated primary tumors. TGF alpha is the so-called growth factor that binds to the EGF receptor, so they belong to the same family of genes.

DR. JÄNISCH: Dr. Kleihues mentioned briefly his very interesting case of a child who had a medulloblastoma, or medulloblastoma-like tumor, which later on became, as he called it, a "gangliocytoma". We had, as I mentioned previously, a similar case published a few years ago, for which our final diagnosis was ganglioglioma because we saw neoplastic astrocytes together with neoplastic ganglion cells. This indicated to me that there was differentiation along two lines, one astrocytic and the other neuronal. Presumably, if you called this tumor *gangliocytoma*, you did not see any neoplastic astrocytes.

I think all of these cases and a few more published in the literature, are of utmost interest in the theoretical and maybe even in the practical sense. It would be useful if we knew under what circumstances these tumors become differentiated and almost stop growing.

DR. HASSOUN: I would like to make a short comment on the paper of Paul Kleihues concerning neurocytomas which developed in the midbrain of adults. The very clear demonstration of synaptophysin in these tumors corresponds to the electron microscopic demonstration of more or less complete synapses. These tumors must be considered as highly differentiated neuronal tumors, not made up of ganglion cells but small round neuron tumors.

We now have collected 14 cases of such tumors, employing stereotactic biopsy, which we have found to be very useful. It is very interesting to note that by immunocytochemistry these tumors are neurofilament negative (210 kD, 160 kD,and 70 kD neurofilaments). Such criteria of neuronal differentiation are absent in these otherwise highly differentiated tumors.

DR. KLEIHUES: Your data are largely confirmed by our own observations. In only one out of approximately 8 cases of central neurocytoma did we find focal expression of neurofilament protein and this was associated with the appearance of clusters of ganglioid cells. However, all neurocytomas showed diffuse expression of synaptophysin, e.g. an antigen localized in presynaptic vesicles. This is very much in accordance with your electron microscopic studies.

DR. HASSOUN: We must be very cautious when we speak of differentiation. There are many levels of differentiation depending on the cell type taken for reference.

DR. KLEIHUES: A round cell tumor like the neurocytoma, which has synaptophysin already present, is already at a pretty high level.

DR. HASSOUN: The small round cells can be the end of the process of differentiation towards a certain shape or type of neuron. The ganglion cell is not the only result of neuronal differentiation and is not the only type of neuron in the organism.

DR. SZYMAS: May I expand our discussion to include another marker because I think this is the first one which provides differentiation between malignancy and brain tissue (SYZMAS et al. 1987a). Such a marker is the monoclonal antibody against the C-15 cluster, commercially known as Leu-Ml. This is usually used as a marker for monocytes and macrophages and as a differentiation marker which also recognizes some of the stem cells. This antibody gives a very strong reaction with brain tissue. Brain tumor is negative. The border is very sharp, and reactive brain tissue has a very striking positivity. In immunoelectron microscopy, one can see that this positivity is connected with the extracellular space (SZYMAS et al. 1988b) and sometimes with the membrane of the astrocytes.

We think that lacto-III-pentose, the antigen which is recognized by this antibody, is connected with glycoproteins present in the extracellular matrix. This glycoprotein is produced by astrocytes and after neoplastic transformation these cells lose the ability to produce this antigen.

I think it is a very useful marker for the differentiation between malignancy of brain tissue and brain tissue itself.

CHAPTER X

Monstrocellular Sarcomas

KLAUS J. ZÜLCH, *Max-Planck Institute for Brain Research, Cologne-Merheim, Federal Republic of Germany*

I would like to review for you one of my special interests, monstrocellular sarcoma, a tumor entity which appears in the WHO classification under two terms: (a) giant cell glioblastoma or (b) monstrocellular type of sarcoma. In the older literature this tumor has been variously referred to as: gangliogliioneuroma, ganglioneuroma, spongioblastoma multiforme ganglioides, spongioastroblastoma, spongioneuroblastoma and retothelial sarcoma (reticulosarcoma).

Discussion of the position of monstrocellular sarcoma in the system of intracranial tumors has continued since they were first described by SCHMINCKE (1909). However, they were included in the nerve cell tumors by SCHMINCKE (1914) and in the glioblastomas ("spongioblastomas") by FOERSTER and GAGEL (1931). My own investigation (ZÜLCH 1939), led me to discuss their sarcomatous nature (ZÜLCH, 1940). In 1948 I demonstrated 22 cases and defined them as an entity which was thereafter broadly described under the term of *monstrocellular sarcoma* (ZÜLCH 1953). The epithet *monstrous* was first used by CUSHING and EISENHARDT (1938) in another context.

Some authors are still not convinced of the mesodermal origin of these monstrocellular tumors, interpreting them as *giant cell glioblastomas* or *monstrocellular astrocytomas* as expressed by LYNN et al. (1968). KERNOHAN and UIHLEIN (1962) preferred the term of *giant cell fibrosarcoma*, whereas KUBO et al. (1974) were in favor of the term *monstrocellular sarcoma* since neither electron microscopy nor tissue culture revealed a difference between glial filaments and tonofibrils. After all these discussions and new investigations, I am still of the opinion that the monstrous character of the cells, which sometimes attain a size of 0.5 mm and are thus visible to the naked eye, merits a special designation under this chosen name. The very abundant formation of reticular silver-staining fibers and the frequent origin of these monster cells far distant from the tumor in the marginal capillaries also suggest a mesodermal character. Such an origin is also indicated by its free invasion of the dura and its not infrequent trend to metastasize extracranially. Finally, there is absence of a predilection for age and size which is a characteristic of sarcomas and not of glioblastomas. They are fleshy tumors with a finely tufted, fibrous cut surface looking rather like asbestos. The varigated color of typical glioblastoma is not seen.

Necroses, fatty degeneration and hemorrhagic foci seen so commonly in glioblastomas are not evident and there are none of the fistulous vessels so characteristic of glioblastomas. Particularly impressive is the fact that mitotic figures may also be grotesquely abnormal. These tumors are frequently cystic with multiple small cysts occurring in one-third of our cases. Their consistency is often tough and hard, due to the abundance of reticulin fibers. They may infiltrate and rupture the dura (infiltration noted in 10 of 71 cases and penetration in 4). Infiltration of the leptomeninges is common. All of the foregoing features are very uncommon in glioblastoma multiforme.

It is stated in the literature that GFAP *markers* may be positive but SZYMAS (personal communication) and others have doubted their specificity. We have never seen calcification in these tumors. Vascular overgrowth, which is so prominent in glioblastomas, particularly at the border of necrotic areas, is much less prominent. The monstrocellular sarcomas seem to have a tendency to diffusely spread into the leptomeninges, a trend which is also uncharacteristic of glioblastomas, although it can occur.

As far as we can tell from observations over more than 30 years, this is a highly malignant, rapidly growing tumor (grade IV in the WHO classification), but with a somewhat better prognosis in some instances. BINGAS (1964) found one case in our series with a 7-year survival.

CHAPTER XI

Grading of Gliomas

RICHARD L. DAVIS, *Department of Pathology, University of California School of Medicine, San Francisco, California*

When Dr. Fields called some weeks ago to ask me to come to this conference and speak on the classification of gliomas, I did not have time to ponder on the request. However, during the ensuing days, as I thought about my presentation, I began to grasp the difficult task I had accepted. To talk on such a subject to this audience, most of whom have spent a large part of their professional lives grappling with brain tumors, is no mean undertaking.

It is impossible to do this subject justice in a short period of time. I will, therefore, present my abbreviated version of the history of glioma classification (my version of the "truth" if you will), and spend a bit more time on the main classifications used by four of the larger centers actively accruing cases for the study of brain tumor therapy. This will permit examination of the major differences that exist between these centers as to the definitions of the various types of glioma, or more specifically astrocytoma-glioblastoma, and also highlight the strong similarities between them.

As an aside, I will not attempt to deal with the oligodendrogliomas, since the schemata for their classification are too new and untested "in the field" for us to make meaningful statements about their utility yet.

So, what is the history of astrocytoma-glioblastoma classification? I believe that the modern history of such classification begins with BAILEY and CUSHING (1926) as put together, I hear, mainly by Cushing. This classification of CNS glia, with primitive, bipolar, and unipolar spongioblasts, as well as astroblasts, and fibrillary and protoplasmic astrocytes, resulted in a tumor classification including spongioblastoma multiforme and unipolare, astroblastoma, and protoplasmic and fibrillary astrocytoma. As Brucher, I think correctly, commented this scheme was "too sophisticated" for many neurologists and neurosurgeons (at least of that time), and when the KERNOHAN et al. (1949) proposal emerged, it was greeted with much favor.

The KERNOHAN and SAYRE (1952) proposal, presented in complete form in the fascicle co-published by the Armed Forces Institute of Pathology, divided the astrocytomas into four *grades*, with glioblastoma multiforme being grades III and IV.

I think there is an important aside here. Many do not appreciate the origin of the four grades; BRODERS (1926) first used these for the classification of skin and bladder cancers, and decided against the well-known 3-tiered system in favor of a 4-tiered system. He did this *by dividing the highest, grade III, into two* thus making the old highest grade into grades III and IV. Kernohan, following this example, did the same and as the fascicle clearly states, grade III and IV astrocytomas are equivalent to glioblastoma multiforme, being divided by degrees within this already fearsome lesion.

The result of these manipulations was to make the grade II astrocytoma cover a broad range of tumors, with widely variant morphology, with a ready aptitude for misinterpretation. This whole difficult area was further complicated by the fact that *all* astrocytomas were included in the grading plan, including tumors, such as the common cerebellar astrocytoma, with little or no tendency to infiltrate or to become anaplastic with time. Though the goal was laudable, it added further to the confusion that followed, and unfortunately delayed a broader understanding of the biology of brain tumors of astrocytic lineage.

As STAGE and STEIN (1974) discovered, at least the way the grading system was being used, it did not reflect very well the biology or response to therapy in a useful way.

I believe that there were some very understandable reasons for this, but not particularly desirable ones. As individual pathologists dealt with brain tumors, they realized the broad range of tumors covered by *grade II* and found it uncomfortably broad. Therefore, without any definitions or agreememt, there were shifts in diagnosis, not consistent from one area or hospital to another, particularly from the high end of grade II into grade III. Because there were no agreed upon definitions, there was little consistency in diagnosis, and the result was minor chaos. Diagnoses of grade I-II or III-IV or even more often of II-III became common. I can attest that this still persists, as I commonly get cases with such hyphenated diagnoses,, and even in the extreme diagnoses such as grade I-IV!

And, as alluded to above, since there was nothing to intellectually separate the highly infiltrative tumors of type *grade II* from some of the circumscribed cerebellar tumors also being assigned this grade, the system was am impediment to further general understanding of the actual differences in growth behavior and potential.

As a result of this, there was movement away from the Kernohan system to more complex systems (our colleagues seemed ready for some sophistication now). Such systems, however, still had to deal with the morphobiologic spectrum of the infiltrating astrocytic neoplasms, often culminating in glioblastoma; they did this in a variety of ways.

The threads lead from BAILEY and CUSHING (1926), influenced by PENFIELD (1932), RINGERTZ (1950) and KERNOHAN and SAYRE (1952), to Dorothy Russell and her many students, and on through the many contributing continental pathologists up to recent times.

After debates and discussions, some of which, I have been told, were rather heated, there emerged the WHO classification, a system that was released in the late seventies (ZÜLCH 1979) after many years of painful labor.

I must say that I was among those applauding loudest when the WHO report was issued. However, as I read it with care and had experience trying to utilize it, I had to read it again, realizing that some aspects were difficult to apply with any reality, and some designations just did not make sense to me. One of these is the use of the term *benign* as related to glial neoplasms. I find that my experience tells me that this is a term that should not be used to describe such tumors, for it conveys incorrect information to both patient and physician, and promises what is likely not to be delivered. I had further misgivings when I was told that gemistocytic astrocytomas, tumors I had learned to fear for their growth potential, were "held to correspond histologically to grade II". Further, I found the assignment of grades to brain tumors as suggested by the report to be opaque and of little use.

Others too shared these misgivings, and to support several large brain tumor studies there were developed similar though not identical systems of classification (KRAMER 1976; LAERUM et al. 1984; PIEPMEIER 1987; LINDEGAARD et al. 1987; AMMIRATI et al. 1987). I have chosen four of these to represent examples of the differences and similarities.

The center utilizing the diagnostic classification closest to that proposed by the WHO is, as far as I can tell, the EORTC Brain Tumor Cooperative Group (BRUCHER et al. 1987), with the *referee* function centered in Brussels under the direction of Professor J-M Brucher who is present at this meeting and will be able to comment and enlarge upon my brief summary of their work.

In their paper, there is a schematic representation of the four-grade grouping as used in their studies. Their *grade I* suggests a delimited lesion. Does this group include such delimited lesions as juvenile pilocytic astrocytoma? Or are these potentially infiltrating tumors just at their inception? If the former, then I cannot agree to their inclusion with the infiltrating tumors, for there are only very rarely transitions, and this would represent another oversimplification reverting to the Kernohan scheme. If the latter, then this is a tumor type so rare as to be beyond my experience, and I would like to see examples.

In my experience very *low grade* infiltrating tumors are just that, clearly infiltrative and without discernable borders. If there is need to indicate anaplasia in otherwise circumscribed tumors, then using the adjective *anaplastic* would seem to me to satisfy this need. To group rarely anaplastic, usually mass-increasing, non-infiltrative tumors with those frequently showing these features is to me retrogressive. Now, though I think I understand what is meant by grades II and III, I must register here my concern about the use of the terms *benign* and *semi-benign* for infiltrative tumors. What I call tumors of this category are potentially lethal lesions, even though they take 3 or 5 or 10 years to kill. They often evolve into more malignant tumors, i.e., they become more anaplastic with time, and they are not amenable to complete removal by any usual means. To me, this would be comparable to calling a malignant melanoma that happens not to recur and kill for 10 years an *benign* tumor. Nonsense!

Finally, as I understand it, the *grade III* and *grade IV* lesions are not separable histologically by their intrinsic features, but only by the fact that one shows focal anaplasia in a more diffuse lesion, the other showing a relatively sharp border between anaplastic areas and relatively normal brain tissue. To me this too is retrogressive, harking back to the Kernohan grouping of grades III and IV as *glioblastoma multiforme*. If this is so, then again, grade II has the potential for extreme variation, and we are back to the potential errors and zone of uncertainty that was the fatal flaw of the Mayo grading scheme. For me, this is one of the most cogent reasons for the abandonment of the term *grade* for long enough for it to lose all connection with that system. Until it does, the use of a *grade* cannot help but be contaminated by that association, no matter how distant.

Also in their paper there are two figures demonstrating a clear separation between grade III and IV tumors with and without adjuvant chemotherapy, both with statistical significance. Again, this is to be expected with consistently applied criteria within a closely knit diagnostic group. If the group does not agree with itself, then indeed there is trouble.

Finally, they have a table that shows the well known difference in survival depending upon age. Malignant tumors in older people have a much worse prognosis than less anaplastic tumors in younger people. All the studies to be shown today underscore that phenomenon. No data are given for the *grade I* or *II* lesions, so that direct comparison with some of the later work with these lesions is not possible.

The second group whose work I will summarize is that at Duke University. It is helpful that Dr. Peter Burger is among us at this meeting to add his intimate knowledge of this work.

The Duke group's paper (Burger et al. 1985) on these tumors shows the histologic criteria used in their assessment of the tumors. It begins with the lower grade tumors, as did that of the Brussels' group. In their scheme necrosis is *required* for the diagnosis of glioblastoma multiforme, but vascular proliferation (type not specified) is not. Further, necrosis is not acceptable in the group of anaplastic astrocytomas. It is not stated what would be done with a lesion *with* necrosis, but otherwise with only the features of astrocytoma. Again, with this group the survival curves show that they are internally consistent, and this is to be expected. Moreover, the effect of age and increasing anaplasia on survival is again well documented. Burger et al. (1985) also did individual feature correlations.

Next I wish to present the results and criteria used by the St. Louis group (Nelson et al. 1983). It is unfortunate that neither Dr. Nelson nor a representative of this group could be with us to add to our understanding of their results. In their article they describe the criteria that they used to separate *astrocytoma*, *AAF* (astrocytoma with atypical and/or anaplastic features), and *GBM* (glioblastoma multiforme). Unfortunately, the definitions which they use do not match exactly those of any other investigators. In the AAF group vascular *prominence* is required and no tumor necrosis is permitted; however, necrosis is *required* for GBM, as are all the other features of AAF. In another paper they showed that necrosis was a significant feature in predicting outcome in these tumors, but since that is a definitional criterion for a tumor with known malignant behavior, can it be used both as a criterion and a tested variable? In San Francisco, we also had a significant difference in survival between our GM and HAA, but we do not require necrosis for the diagnosis of GM! We are independently testing this variable in our ongoing studies.

I would like now to briefly summarize the work and criteria that we have used in San Francisco (UCSF)in our studies during recent years (Table I). We have employed the following definitions in our diagnosis of these infiltrating brain tumors. Our scheme requires that one begin with the criteria for the most malignant tumor and move down until the criteria are met, and then stopping with this diagnosis. This may be significantly different from other schemata but we think it works well. Furthermore, we do not require necrosis for the diagnosis of *glioblastoma multiforme (GM)*. We have too often not seen necrosis in the biopsy specimen and then found it in abundance in the brain tumor seen a few days or weeks later. Thus, we felt we would be under-diagnosing this tumor if necrosis were a required feature.

TABLE I Diagnostic criteria used at UCSF for the diagnosis of cerebral glioblastoma and infiltrating astrocytomas

Glioblastoma Multiforme (GM)

1. At least focally highly cellular glial neoplasm, and
2. Nuclear pleomorphism, and
3. Cytoplasmic pleomorphism, and
4. Vascular endothelial proliferation

Gemistocytic Astrocytoma (GA)

1. Not a glioblastoma multiforme.
2. An astrocytoma composed of at least 60% gemistocytes

Highly Anaplastic Astrocytoma (HAA) (Infiltrating)

1. Not a glioblastoma multiforme or gemistocytic astrocytoma, and
2. At least focally moderately to highly cellular, and
3. High N/C ratio, or
4. Coarse nuclear chromatin, or
5. Much mitotic activity, or
6. Nuclear pleomorphism, or
7. Cytoplasmic pleomorphism

(Need at least 2 of items 3 to 7)

Moderately Anaplastic Astrocytoma (MoAA) (Infiltrating)

1. Not a highly anaplastic astrocytoma, and
2. Mild to moderate increased cellularity, and
3. Enlarged nuclei, and
4. Relatively uniform cytoplasm

Mildly Anaplastic Astrocytoma (MiAA) (Infiltrating)

1. Not a moderately anaplastic astrocytoma, and
2. Mild increase in cellularity, and
3. Nuclei enlarged, but regular, and
4. Uniform cytoplasm but not gemistocytic

TABLE II Comparison of systems of classification

	DUKE (BURGER et al.1985)	WASH. UNIV. (NELSON et al.1983)	UCSF
Glioblastoma Multiforme (GM)			
Cellularity	mod to marked	increased	highly cellular, at least focally
Nuclear pleomorphism	mod to marked	multifocal or diffuse	present
Cellular pleomorphism	mod to marked	multifocal or diffuse	present
Vascular proliferation	OK, not req	prominence	endothelial proliferation
Necrosis	req, with or w/o pseudopalisading	one or more foci	OK, but not req
Other	mitotic figures		
"Malignant Astrocytoma"			
	Anaplastic	**"AAF"**	**Highly anaplastic astrocytoma (HAA)**
Cellularity	moderate	increased	moderate to high
Nuclear pleomorphism	moderate	multifocal or diffuse	OK, not req
Cytoplasmic pleomorphism	moderate	multifocal or diffuse	OK, not req
Vascular proliferation	OK, not req	"prominence"	OK, not req
Mitotic figures	not stated	"increased"	OK, not req
Other	no necrosis	no tumor necrosis	OK, may not meet criteria for GM (see Table I)
"Less Anaplastic Astrocytoma"			
	Astrocytoma	**Astrocytoma**	**Mod Anaplastic Astrocytoma (MOAA)**
Cellularity	"slight hypercellularity"	moderate	mild to mod hypercellularity

(continued)

TABLE II (continued)

	DUKE	WASH. UNIV.	UCSF
Nuclear pleomorphism	slight	"uniform cells closely resembling mature, resting or reactive non-neoplastic astrocytes"	increased
Cytoplasmic pleomorphism	slight		
Vascular proliferation	none	not stated	(usually none)
Necrosis	none	(none)	(usually none)
Other	mitoses absent or very rare		

Note that this system is basically a three-tiered one, with glioblastoma at the top, gemistocytic astrocytoma and highly anaplastic astrocytoma the next group, and moderately anaplastic astrocytoma the lowest of the three. Gemistocytic astrocytoma was separated because of its morphologic uniqueness, though we appreciated that these tumors might well be a part of either GM or HAA. As it turns out, they belong with the HAAs. And the MiAAs are so rare, only a small handful of cases being seen in a year, that they, for practical purposes can be grouped with the MoAAs.

Our survival statistics, using a Kaplan-Meyer curve, also demonstrate good separation of the three categories. It is interesting that all three of the last groups have happened on a three-tiered system independently, and that the statistics hold up for each, even though the detailed criteria, also arrived at independently, are subtly different.

It is more important to note, that we have included survival statistics for the lower grade tumors, the MoAAs. These show very clearly that these are not *benign* tumors, though in some grading schemes they would be called *grade II*. They kill, albeit more slowly than their more anaplastic counterparts, which is why we use the term *moderately* anaplastic for these lesions.

It is worthy of independent note, that none of the latter three studies includes (as far as I can tell) any but infiltrating tumors, and do not include such lesions as the juvenile pilocytic astrocytoma, even when occurring in the cerebrum.

Where should we go from here? Are we ready for another round of discussion to again attempt to gain acceptance of a more tightly defined series of gradations or indicators of varying degrees of anaplasia in astrocytic neoplasms and glioblastoma? Or is the time not yet ripe for that development. I don't think that we are quite ready for the next step, myself, but I am willing to try. It will take great patience and thoughtfulness, and a large fund of good will, as well as *an appreciation that others with a differing approach are not wrong, but only have another view of the problem.* Further, though we are all devotees of the morphologic method, we must be ready to accept and utilize new methods to prod and assist us in our future efforts. We have embraced the new morphology of cell markers, and turned them to our benefit. We should be ready to add the technology of labelling index or chromosomal abnormalities and other markers of cell-growth potential to our armamentarium as well.

CHAPTER XII

A New Uniform Grading System (Using Mayo Clinic Material)

CATHERINE DAUMAS-DUPORT, *Service of Pathological Anatomy, Hôpital Sainte Anne, Paris, France*

Introduction

Brain tumors are subdivided into groups defined by their histology. This is true regardless of the viewpoint from which they are studied: surgical, clinical, pathologic, radiologic or in fundamental research. However, if the histologic groups are heterogeneous and not reproducible, a bias is introduced that will dramatically impair the validity of any result determined by any of these disciplines. This should make one conscious of the role and responsibilities of the pathologist in the field of brain tumors.

The WHO proposed a histogenic classification of gliomas and rejected the principle of assigning a numerical grade of malignancy to a given cytologic type of glioma (ZÜLCH 1979). Nevertheless, the grading scheme of Kernohan which subdivides astrocytomas into four grades is still in use (KERNOHAN 1949). The situation has become even more complicated by the recent introduction of 3-tiered grading systems (NELSON et al. 1983; BURGER et al. 1985). The reproducibility of any of these systems has never been tested. What is the chance that one pathologist will classify a given tumor in the same way as others?

The literature is replete with contradictory results concerning the efficacy of the various modes of treatment of gliomas. For example, adjuvant radiotherapy in low grade gliomas has been demonstrated to be either of benefit (WEIR and GRACE 1976; FASEKAS 1977) or ineffective (GARCIA et al. 1985). Similar observations have been reported regarding the efficacy of radical surgical resection of astrocytomas, some series showing prolonged survival (WEIR and GRACE 1976; FASEKAS 1976; LAWS et al. 1984) and others no effect (SCANLON and TAYLOR 1979).

The search for effective therapeutic modalities seems to justify large clinical trials with aggressive radiotherapy-chemotherapy combinations. However, large series of patients can probably not compensate for the bias created by analyzing heterogeneous and/or non-reproducible tumor groups. Thus it is likely that pathologists are partly responsible for the current multiplication of randomized therapy protocols for gliomas.

We have recently demonstrated that homogeneous and reproducible groups of clear-cut prognostic significance can easily be determined in astrocytic tumors by applying a previously published grading system of gliomas (DAUMAS-DUPORT et al. 1982; DAUMAS-DUPORT et al. 1988). The proposed method which utilizes the WHO histogenic classification as well as a numerical grading scheme, will be described in the first part of this article.

Histologic characterization of gliomas has traditionally been restricted to the definition of their histologic type and degree of malignancy. However, modern imaging techniques and increasing applications of stereotactic methods in the diagnosis and in the treatment of brain tumors are opening new approaches to the histologic assessment of glial neoplasms. Nowadays, precise information can be obtained safely *in vivo* regarding the extent as well as the structure of the tumor, i.e., on the presence of solid tumor tissue, infiltrated parenchyma, or both. This led us to propose a classification scheme of gliomas according to three distinct structure types. This classification has been designed for utilization in addition to the histologic type and degree of malignancy (DAUMAS-DUPORT and SZIKLA 1981; DAUMAS-DUPORT et al. 1982).

Although this simple concept has primarily been derived from stereotactic methodology, study of the structure type of gliomas may also prove useful in conventional surgical pathology practice. Contribution of the structure type to the determination of homogeneous tumor groups as well as for establishing a rational basis for the treatment in individual patients harboring gliomas will be detailed in the second part of this article.

Grading System of Astrocytomas

In 1981 a simple grading system was proposed as a two-part classification scheme of gliomas, the two features being: a) histologic malignancy and b) structure of the tumor (SCANLON and TAYLOR 1979).

The grading system for gliomas relies upon recognition of the presence or absence of four parameters: nuclear atypia, mitosis, necrosis, and endothelial proliferation. The method is binary and results in a summary score which is translated into grades as follows: zero criterion = grade I, one criterion = grade II, two criteria = grade III, three or four criteria = grade IV.

A recent study (DAUMAS-DUPORT et al. 1982) allowed us to demonstrate the effectiveness and reproducibility of this grading system in *ordinary* astrocytomas, i.e., fibrillary, protoplasmic, gemistocytic and anaplastic astrocytomas, as well as glioblastomas. The histologic material and clinical data upon which this study was based were derived from a previously reported series of patients with astrocytomas, radiotherapeutically treated at the Mayo Clinic between the years 1960 and 1969 (DAUMAS-DUPORT and SZIKLA 1981). a 15-year period of follow-up was available on all surviving patients. From this series, initially graded I to IV according to the Kernohan system in which no distinction was made between the various histologic subtypes of astrocytomas, 287 ordinary astrocytomas were entered into the study, 51 pilocytic and microcystic cerebellar astrocytomas were also included for the purpose of comparison.

Statistical analysis showed that in ordinary astrocytomas, grades were strongly correlated with survival ($p<0.0001$). The median survival was 4 years in grade II, 1.6 years in grade III, and 0.7 years in grade IV tumors. Of the only two patients with grade I ordinary astrocytoma, one survived 11 years, the other 15 years. Furthermore, based upon the Cox model, grade was found to be the major prognostic factor superseding the effects of age, sex, and location.

Double-blind grading between two independent observers was concordant in 94% of ordinary astrocytomas. Reproducibility was 81% in low grade (grades I and II) and 96% in high grade (grades III and IV) tumors. These satisfactory reproducibility rates are attributed to the selection of easily-identified morphologic criteria and to their simple recognition as present or absent. Application of this simple and reproducible method for the study of ordinary forms of astrocytoma should permit reliable comparison of clinical and therapeutic data emerging from various treatment centers. The reproducibility of the proposed method of grading is in fact the most convincing argument in favor of its adoption.

Since the validity and reproducibility of any grading system depends on the homogeneity of the lesions under consideration, it is no surprise that strict definition of histologic subtypes of astrocytomas is a major element determining the efficiency of our proposed method. The behavioral characteristics of pilocytic and cerebellar-type astrocytomas prompted their consideration as separate entities (KERNOHAN 1949; RUBINSTEIN 1972a; WEIR and GRACE 1976; FASEKAS 1977; ZÜLCH 1979; SCANLON and TAYLOR 1979; DAUMAS-DUPORT and SZIKLA 1981; DAUMAS-DUPORT et. al. 1982; NELSON et al. 1983; LAWS et al. 1984; BURGER et al. 1985; GARCIA et al. 1985; DAUMAS-DUPORT et. al. 1988). In our study (DAUMAS-DUPORT et al. 1988) a strong association between cell types and survival was noted ($p<0.0001$). Patients with ordinary astrocytomas had a median survival of 0.87 years, less than 5% of the patients being alive at 15 years. Of those with pilocytic or microcystic cerebellar-type astrocytomas, 15-year survival was

67% and 91%, respectively. Attempts to apply our grading system to these tumors showed no correlation between histologic grade and patient survival.

In addition to their distinct natural history as compared with that of ordinary astrocytomas, pilocytic and cerebellar-type astrocytomas occur primarily in younger patients and arise at characteristic sites. We have shown that mixing of pilocytic and microcystic cerebellar astrocytomas with ordinary astrocytoma cell types introduces a bias which dramatically modifies expected survival in low grade astrocytomas (DAUMAS-DUPORT et al. 1988; SHAW et al. 1989). Therefore, contradictory results in evaluating the influence of adjuvant postoperative radiotherapy (WEIR and GRACE 1976; FASEKAS 1977; GARCIA et al. 1985) or of the effectiveness of radical surgical resection (WEIR and GRACE 1976; FASEKAS 1977; SCANLON and TAYLOR 1979; LAWS et al. 1984), may in large part be explained by heterogeneity in the age of the populations under study, with either more or fewer children being included according to the recruitment pattern of the various treatment centers. Application of our method in a retrospective study of supratentorial low grade astrocytomas (SHAW et al. 1989) has demonstrated that pilocytic astrocytomas are cured by surgical removal and that adjuvant radiotherapy is of benefit only in low grade astrocytomas or in pilocytic astrocytomas which are incompletely resected. Indiscriminate lumping of histologic subtypes also explains the observation that age has an overwhelming effect onpatient survival, eclipsing other variables including histologic grade (STAGE and STEIN 1974; SCANLON and TAYLOR 1979; LAWS et al. 1984).

Conversely, our grading method permits an indirect control in the determination of the histologic type of glioma. In ordinary astrocytomas, the four basic criteria tend to appear in a predictable sequence corresponding to the increase in biologic malignancy (DAUMAS-DUPORT et al. 1988): nuclear atypia, the first criterion, has been shown to be present in 100% of grade II tumors; mitotic activity, the second criterion, was observed in 92% of grade III tumors; necrosis and/or endothelial proliferation were almost exclusively present in grade IV astrocytomas. Thus, for example, in tumors having mitoses but demonstrating no nuclear atypia, or one in which there is endothelial proliferation but no mitosis, a diagnosis of ordinary astrocytoma must be reconsidered in order to avoid confusion between it and other gliomas such as pilocytic astrocytoma or oligodendroglioma. Similarly, in view of the rarity of grade I ordinary astrocytomas, a tumor having no nuclear abnormality, as well as none of the other criteria, should be carefully reexamined in order to avoid misdiagnosis with pilocytic astrocytomas or with reactive gliosis.

Figure 1 Gemistocytic astrocytoma: A, tumor tissue; B, isolated tumor cells (ITCs) in white matter. In the tumor tissue, cells are crowded and are associated with neovascularity (A1 and A2); Bodian stain demonstrates only some residual axons (A3 arrowhead). ITCs have distinct nuclear abnormalities (B1 and B2 arrows) and lack visible cytoplasm, the appearance being that of "naked nuclei". As demonstrated on Bodian stain (B3), these cells permeate an architecturally intact parenchyma. Axons are also readily visible on smear preparation (B2). ITCs are not accompanied by neovascularity. A1 and B1: glutaraldehyde fixation, hemalum-phloxine-safranin stain (x 200). A2 and B2: alcohol-fixed, hemalum-phloxine-stained smear preparations (x 200) A3 and B3: Bodian-hemalum stain (x 200).

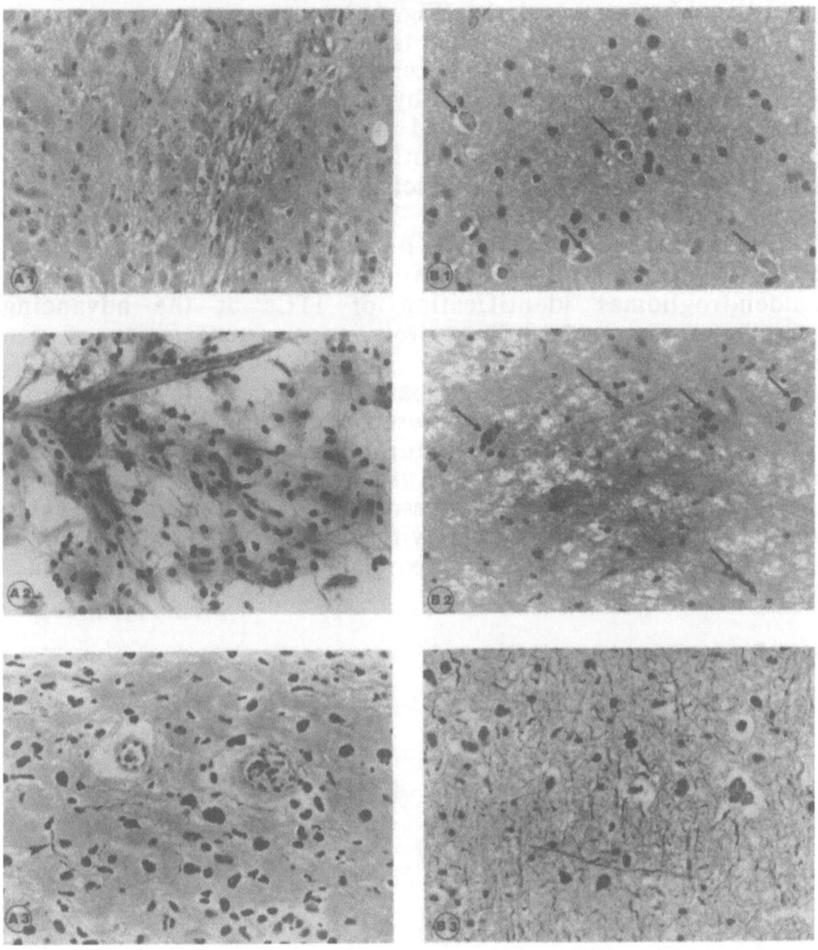

Morphologic Type of Gliomas

Glial neoplasms may be viewed as demonstrating two distinct growth patterns (DAUMAS-DUPORT et al. 1982; DAUMAS-DUPORT et al. 1987): 1) Tumor tissue proper in which the cells are crowded, unaccompanied by intervening normal brain parenchyma and associated with newly-formed blood vessels (Fig. 1A); 2) Isolated Tumor Cells (ITCs), an arrangement wherein the tumor cells are not in contact with each other but permeate largely intact brain parenchyma. This pattern is not accompanied by neovascularity (Fig. 1B).

In astrocytomas, the tumor tissue is made of cells which possess visible cytoplasm and cell process (Fig. 1A), whereas ITCs are usually characterized by a lack of visible cytoplasm giving the appearance of "naked nuclei" (Fig. 1B). Such individual tumor cells can be identified because of distinct nuclear abnormalities. Their lack of visible cytoplasm also permits their distinction from reactive astrocytes, the latter having delicate stellate processes and being evenly spaced in a mosaic pattern (DAUMAS-DUPORT et al. 1987). In oligodendrogliomas identification of ITCs at the advancing edge of the tumor depends entirely on the recognition of nuclear abnormalities.

The two distinct growth patterns, singly or in combination, are regularily encountered in gliomas and are the basis for the following structural classification of gliomas (DAUMAS-DUPORT et al. 1982; DAUMAS-DUPORT et al. 1987), (Fig. 2): Type I - Solid tumor tissue only, no peripheral ITCs; Type II - Solid tumor tissue with peripheral ITCs; and Type III - ITCs within intact brain parenchyma and no solid tumor tissue.

Figure 2 Structure type of gliomas.

As demonstrated in Tables I and II, large variation in structure type are observed among gliomas according to their histologic type and location. If one excludes pilocytic astrocytomas, structure Type I is uncommon among gliomas. Ordinary astrocytomas and oligoastrocytomas most often adopt a structure Type II, while oligodendrogliomas are the most frequent tumors among Type III gliomas (Table I). In series of patients who underwent serial stereotactic biopsies, we also observed that a large majority of gliomas situated on the midline are of structure Type I (80%), whereas this structure represents only 16% of gliomas considered as a group, independently of their localization (see Table II).

TABLE I Structure type of 284 gliomas as determined by serial stereotactic biopsies

	Type I	Type II	Type III
Pilocytic astrocytomas N = 22 = 8%	100%	0%	0%
Ordinary astrocytomas N = 137 = 48%	11%	79%	10%
Oligoastrocytomas N = 7 = 24%	4%	82%	14%
Oligodendrocytomas N = 57 = 20%	9%	50%	41%
All cell types N = 280	17%	67%	16%

TABLE II Structure type of 284 gliomas explored by serial stereotactic biopsies

Structure	Type I	Type II	Type III
All locations	17%	67%	16%
Midline	80%	17%	3%

Structure and Tumor Volume: Because of the increasing applications of stereotactic neurosurgical techniques, not only in the diagnosis but also in the treatment of brain tumors, a demand regarding the spatial delimitation of gliomas is being created. Definition of the tumor volume must include not only the solid tumor volume but also those areas of parenchyma infiltrated by ITCs. Underestimation of ITCs is no doubt the basis of a recurrence after a so-called *complete surgical resection*, and will as well be responsible for the failure of more sophisticated "local" destructive treatment such as radiosurgery or interstitial irradiation.

Histologic examination has traditionally been considered an unreliable method for determining the boundaries of gliomas (KLEIHUES and VOLK 1984). In fact, routine histologic techniques normally used for surgical specimen are inadequate for identifying minimum infiltrate within otherwise intact brain parenchyma (DAUMAS-DUPORT et al. 1987).

By applying improved routine techniques for biopsy specimen preparation that simply involve the use of glutaraldehyde-fixed and hemalun-phloxine stained paraffin sections, as well as alcohol-fixed hemalum-phloxine stained smears, we have shown that, in concert, these methods allow visualization of delicate cytologic details which are complementary in assessing the presence of ITCs at the periphery of gliomas (DAUMAS-DUPORT et al. 1987).

Figure 3 Structure type III oligodendrocytoma. Portion of same sample fixed A) in glutaraldehyde and B) in formalin demonstrating that, with formalin fixation, oligodendrocytes infiltrating white matter may be misinterpreted as fibrillary astrocytoma. C) Bodian stain of the formalin fixed portion showing that the fibrillary background is made of axons. (A&B x 200, C x 250).

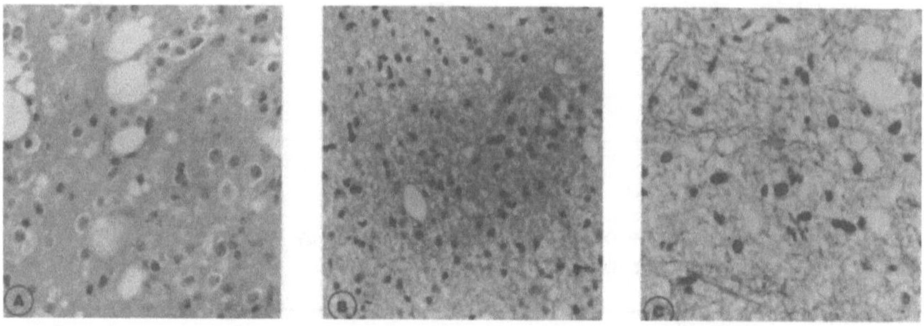

Structure and Identification of Histologic Types of Gliomas: Since the structure of gliomas is not considered in conventional surgical pathology practice, pathologists usually pay no attention to identifying axons intermingled within glial neoplasms.

However, because of frequent artifactual distortion created by careless handling of fresh tissue, and of tissue retraction induced by formalin fixation, entrapped neurons may easily mimic a fibrillary astrocytic background. As a consequence infiltrative oligodendrocytomas of structure Type III are commonly misinterpreted as fibrillary astrocytomas (Fig. 4).

Figure 4 Structure type III oligodendrocytoma. A) After formalin fixation, tumors with such appearance are often misinterpreted as fibrillary astrocytoma. B) BFAP immunostaining demonstrating only scattered reactive astrocytes. C) Bodian stain showing that, in fact, the fibrillary frame is made of resident axons. D) Isolated oligodendrocytes and axons as best demonstrated on smear preparation. (A,B&C: formalin fixation, x 200; D: smear preparation x 300.

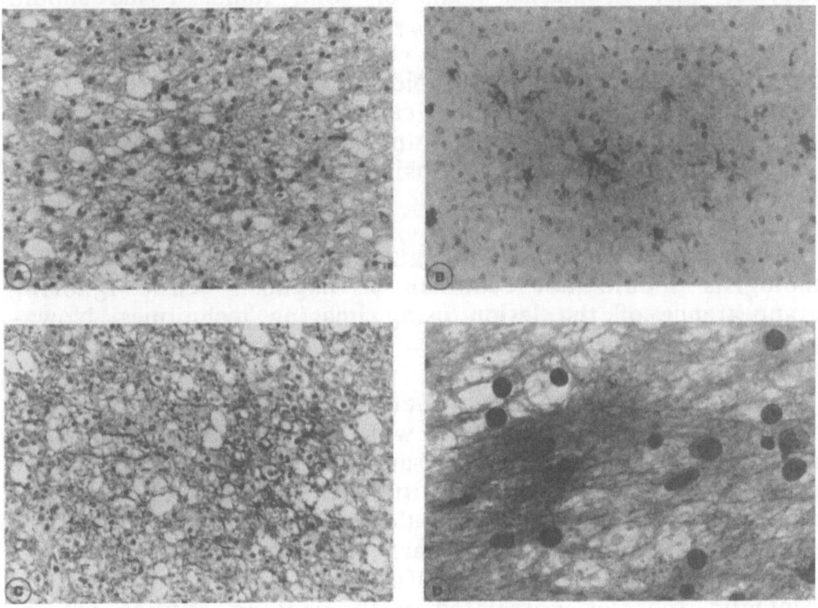

In our experience, improvement of routine techniques, including smear preparations and gluteraldehyde fixation, has dramatically increased the apparent frequency of oligodendrogliomas. These tumors, which represented 5% of gliomas in our earlier studies ((DAUMAS-DUPORT and SZIKLA 1981), have reached up to 20% in our recent series of gliomas diagnosed by serial stereotactic biopsies (Table I). Since isolated tumor astrocytes of Type III astrocytomas are also commonly misinterpreted as fibrillary atrocytomas (DAUMAS-DUPORT et al. 1987), the so-called *fibrillary astrocytomas*, actually correspond to a heterogeneous group of glial neoplasms with very different prognosis.

The structure of gliomas should also be considered in order to make an accurate diagnosis of pilocytic astrocytomas. In our opinion, among astrocytomas consisting of bipolar astrocytes, only neoplasms made of solid tumor tissue should be identified as pilocytic astrocytomas. These tumors should be clearly distinguished from ordinary infiltrative astrocytomas which may adopt a piloid appearance in certain brain structures such as the corpus callosum and brainstem. This distinction is of paramount importance since these tumors have a very different prognosis and require a distinctly different form of treatment.

Structure of Gliomas and Imaging Techniques: In our opinion, pathologists should not establish a diagnosis while ignoring the appearance of the lesion using imaging techniques. Nowadays modern imaging techniques provide the equivalent of a macroscopic examination, which, with the exception of large, *en bloc* removal most often default in the field of brain surgical pathology. Given the well known heterogeneity of glial neoplasms, adequacy of sampling as provided by the neurosurgeon is a crucial point in histologic evaluation. Conversely, the pathologist should be capable of evaluating the validity of samples which are submitted for examination. In this regard the following information derived from serial stereotactic biopsies correlated with imaging studies are of paramount practical importance (DAUMAS-DUPORT et al. 1987; KLEIHUES and VOLK 1984).

It has been demonstrated that there is a strong correlation between the degree of tumor microvascularity and CT contrast enhancement. When present, CT contrast enhancement corresponds to the solid tumor tissue itself and does not occur in areas containing ITCs which are devoid of neovascularity.

Since presence of ITCs is accompanied by edema, parenchyma infiltrated by ITCs usually appears as area of hypodensity on CT scan, or as an area of hypersignal on MRI T2 images (KLEIHUES and VOLK 1984; DAUMAS-DUPORT et al. 1987).

As a consequence, a relationship is observed between the structure type of gliomas and CT images. Contrast enhancement is constant in structure Type I, but absent in structure Type III gliomas. In structure Type II tumor, contrast enhancement is inconstant, depending on the degree of neovascularity (DAUMAS-DUPORT et al. 1987).

Therefore, if no obvious neovascularity is apparent on histologic examination, but contrast enhancement is present on CT scan, it can be deduced that the neoplasm possesses a solid tumor tissue component which has not been sampled. In such cases, if no further sampling can be obtained, the grade of the tumor must not be established on paraffin section alone, the final report should also clearly point out that the samples correspond to the periphery of a glioma for which the histologic subtype cannot be precisely defined.

Structure Type of Gliomas and Therapeutic Strategy: The histologic demarcation of an intracranial neoplasm into tumor tissue proper and infiltrated parenchyma is important therapeutically. Tumor tissue that contains no intact parenchymal elements can be destroyed without a resultant deficit. Parenchyma infiltrated by ITCs is presumably still intact and must be preserved if a neurologic deficit is to be avoided.

Structure Type I gliomas such as pilocytic astrocytomas can be cured by surgery alone; the tumor volume can be estimated from CT contrast enhancement. Precise removal of deep-seated tumors or of neoplasms situated in functionally important brain structures can be achieved by employing a sterotactic surgical method without a resultant deficit (McGIRR et al. 1987), or can be treated by radiosurgery or interstitial irradiation which allows the destruction of a precise target volume.

In structure type II gliomas, removal or selective radiosurgical destruction of the solid tumor tissue proper can theoretically can be achieved in any location. At the present time, external beam irradiation with or without additional chemotherapy is the only treatment available for the infiltrative portion of this type tumor.

In structure Type III tumors, the extent of ITCs usually precludes complete removal or other local destructive therapy. Partial resection of these tumors which occupy intact parenchyma is of questionable benefit. External beam irradiation, with or without additional chemotherapy, seems to be the most reational basis for treatment of Type III gliomas.

Assessment of the extent of tumor infiltration of Type II and Type III gliomas permits an adequate definition of the radiation field. This is particularly important in low grade gliomas for sparing the patient the delayed effect of ineffective extensive radiotherapy.

Today, a precise spatial definition of cerebral gliomas can safely be obtained from image-guided serial stereotactic biopsies. Modern imaging techniques and increasing applications of stereotactic methods are opening a new approach in the diagnosis and treatment of brain tumors. There is little to suggest that stereotactic procedures will other than increase in importance and that a growing number of pathologists will be involved in this practice.

Given the understandable concerns when biopsy sampling is considered, it must, however, be stressed that, in fact, there are very different kinds of so-called *stereotactic biopsies*. Free-hand biopsies done in the CT scanner or single sample CT-based stereotactic biopsies must not be equated with carefully performed stereotactic serial biopsies, based on imaging data, performed by experienced stereotacticians.

Histologic information concerning the spatial definition of gliomas is of paramount importance for planning therapy in individual patients. The potential of these methods will, however, in large measure, depend upon a willing attitude on the part of pathologists. Thus, this schematic survey of the contribution of histologic spatial data in the management of gliomas has been mainly intended to suggest that, for the pathologist, stereotactic methods can open more useful and more exciting perspectives than conventional neurosurgery, for which the role of histopathology has been restricted to define, postoperatively, the histologic type and degree of malignancy of the tumor.

CHAPTER XIII

The Grading of Astrocytomas and Oligodendrogliomas

PETER C. BURGER, *Department of Pathology, Duke Medical Center, Durham, North Carolina*

Introduction

The grading of astrocytomas and oligodendrogliomas persists as a controversial subject that is not likely to resolve spontaneously into a consensus on a single grading system, universally applied. Such an unanimity seems especially unlikely since many of the present histologic classifications *work*, that is, they assign patients to groups whose survival curves are better or worse than patients with lower or higher histologic grades, respectively. Nevertheless, some common, or at least widely used, system is desirable as is apparent in the standardizing attempts of the World Health Organization *Blue Book* in its original ZÜLCH (1979), and soon-to-be-revised form.

Fibrillary or "Diffuse" Astrocytic Neoplasms

The fibrillary or *diffuse* astrocytic neoplasms are the most common gliomas of the cerebral hemispheres in adults, and the brainstem of children. One of its members, the glioblastoma multiforme, is the most common primary brain tumor. In the cerebral hemispheres these neoplasms range from a slowly growing well differentiated neoplasm with a long history of symptoms, to a glioblastoma which characteristically appears more abruptly in an older patient. Between these two extremes is a range of neoplasms with varying, and often mixed, degrees of differentiation.

Properties of Fibrillary Astrocytic Neoplasms and Limitations of Grading: Any discussion of grading astrocytomas must recognize four cardinal properties. The first is their constitution as a continuum of histologic malignancy that extends from the well differentiated astrocytomas at one end to tumors of overt anaplasia, the glioblastoma multiforme, at the other. Although, molecular events may be described ultimately which define a stepwise evolution and divide this spectrum into its elemental bands, this is not the case for subdivisions based on morphologic criteria. Thus, grading astrocytomas into histologic grades at the same time is arbitrary and subjective. How subjective is apparent to anyone who restudies histologic sections in light of his or her diagnoses made several years earlier.

The second important property of the fibrillary astrocytomas is the extremely close interrelationship between (1) the histologic degree of differentiation, (2) four specific clinical variables, and, (3) the length of survival (GEHAN and WALKER 1977; SHAPIRO et al. 1989; BURGER and GREEN 1987). The importance of these interrelationships is that histologic grading must be done in the recognition that a histologic grade is but one of the factors which influence the subsequent course of a disease. These interrelationships were well documented when the Kernohan system was first applied to the material from the Mayo Clinic (KERNOHAN and SAYRE, 1952). It was found that with increasing histologic grade there was a decreasing duration of neurologic symptoms, an increasing age, and a decreasing duration of postoperative survival. Multiple subsequent studies have reaffirmed these close relationships and added the observation that the degree of neurologic impairment, and the time of presentation, is also inversely related to histologic grade and length of survival.

Of the clinical variables that effect the length of survival, it is the age of the patient that repeats as the most powerful prognostic factor. The relative effects of age, duration of symptoms, and the histologic grades were studied by multivariate analysis in one study, and the histologic grade was eliminated as an independently useful prognostic factor (COHADON et al. 1985). In other studies such as those of the Brain Tumor Cooperative Group, the neurologic performance status, the age of the patient, and the histologic grade all have strong predictive power, and each has been felt to be independently statistically significant (GEHAN and WALKER 1977; SHAPIRO et al. 1989). The prognostic effect of age occurs not only when the whole spectrum of astrocytic neoplasms is considered, but is also noted within subcategories such as the glioblastoma multiforme or *anaplastic* astrocytoma. The explanation for this strong relationship is not clear. It has been our experience that glioblastoma multiformes in older patients have more histologic features of *malignancy* than those in younger patients, but a multivariate analysis suggested that the differences in histologic features between younger and older patients accounted for only a small fraction of the differences in survival (BURGER et al. 1987a).

Given the obvious importance of these clinical variables, it is unfortunate that many published accounts of survival following experimental treatments provide limited information about clinical or pathologic variables. Under these circumstances it is difficult to compare histologic grading systems and to compare results of treatment. A new treatment applied to a group of patients with a median age of 45 years will appear more successful than one given to patients with a median age of 50 years.

The third of the four cardinal features of the fibrillary astrocytic neoplasms is the regional variation in degrees of differentiation. This intratumoral heterogeneity is especially evident in the glioblastoma multiforme, but can also be seen in many neoplasms of lesser grades where there are foci of increased cellularity and mitotic activity (SCHERER 1940). In some neoplasms, it seems inescapable that this mixture is a consequence of tumor progression within a previously better differentiated glioma. The regional heterogeneity of astrocytomas has obvious practical consequences in regard to tissue sampling and tumor grading, especially that done on tissue obtained by stereotactic biopsy. Prognostically significant histologic features like vascular proliferation and necrosis may well be absent in a small sample of an otherwise typical glioblastoma multiforme. In this increasingly frequent setting, the derived histologic grade will underestimate the biologic potential of the neoplasm.

The tendency for tumor progression or a decrease in the degree of differentiation with time is the fourth important property of the fibrillary astrocytic neoplasms. While this tendency is not a limitation of grading *per se*, all graders must necessarily be humble and recognize that this unpredictable event may suddenly invalidate their previous efforts. The advance in histologic malignancy with tumors has been best studied by MÜLLER et al. (1977a) who compared the grades of fibrillary astrocytomas at the time of the original resection and at the time of reoperation. Of the initial lesions, 72 were astrocytomas and 65 were anaplastic astrocytomas. At the time of reoperation, 10 of the 72 astrocytomas remained astrocytomas, 40 had become increased in grade to anaplastic astrocytomas, and 22 had become glioblastoma multiformes. Of the 65 anaplastic astrocytomas, 36 remained unchanged and 29 had become glioblastoma multiformes. As might be expected, the intervals between operation were shorter in the higher grade lesions, and the transition between grades occurred more quickly. In spite of these studies, the incidence of a change in grade is not known since patients with well differentiated lesions that remained unchanged may not have been reoperated. At present there are several large interinstitutional studies charting the course of the well differentiated astrocytomas. These efforts may provide a clearer understanding of the frequency and timing of malignant degeneration in the fibrillary astrocytic neoplasms.

Review of Grading Systems of Astrocytic Neoplasms: One of the early grading systems utilized in North America was that of BAILEY and CUSHING (1926). These authors saw similarities between neoplastic glia and the cellular events in sequential phases of neuroembryogenesis. Their classification therefore recognized three grades, the astrocytoma, the astroblastoma, and the spongioblastoma multiforme. While a three-tiered grading system has returned to favor, as has the concept that neoplastic cells regain certain properties of embryonic cells, the Bailey and Cushing system has one insuperable limitation - its intermediate lesion, the astroblastoma, is an extremely rare lesion. As a consequence, there was, for all practical purposes, no category for a lesion that was more malignant than an astrocytoma yet less malignant than the glioblastoma multiforme.

Evolving out of the Mayo Clinic which was the locus of origin of BRODERS' (1926) grading system for carcinomas, was the four-tiered Kernohan classification. This classification was enthusiastically received by many pathologists familiar with the concept of *dedifferentiation*, and with four-tiered grading systems as applied to other organs. This system considered the astrocytic neoplasms as differing grades of astrocytoma (SVIEN et al. 1949; KERNOHAN and SAYRE 1952). As one progressed from the best differentiated (grade I) to the most malignant (grade IV) there was a progressive increase in the degree of macroscopic circumscription and an increase in histologic variables such as pleomorphism, necrosis, mitoses, and vascular proliferation. The system incorporated Broders' concept that the grade of a neoplasm was, in part, determined by the percentage of malignant cells. This conflicts with the current practice of grading neoplasms on their less differentiated area and giving little weight to the overall ratio of poorly to well differentiated elements. Nevertheless, as utilized by the authors the system clearly had utility and there were close relationships between the age of the patient, duration of symptoms, histologic grade, and subsequent course of the disease. It is interesting that the cerebellar astrocytomas were included in the grade I category.

Since its initial application, the Kernohan system has been widely used although a problem persisted in regard to astrocytomas grades III and IV. It was unclear to many observers how these two grades were to be distinguished in practice, and if any purpose was served in doing so. Reservations that this subdivision has significant prognostic utility found support in the study from the Radiation Therapy Oncology Group where the median survivals for astrocytomas grade III and grade IV were compared (NELSON et al. 1983). The diagnoses of referring pathologists were utilized and no significant differences in survival were noted. There is no question that one can identify certain histologic features within the glioblastoma category which portend a shorter survival, and that with sufficiently large numbers of patients it would be possible to subcategorize the glioblastoma multiforme into astrocytoma grades III and IV, or even glioblastoma III, IV, and V. However, the differences in survival would not be great and at present such a subclassification would not alter the clinical management.

A year after the Kernohan system made its first public appearance, a system simpler by one grade was presented by RINGERTZ (1950). In this system, an astrocytoma, a malignant astrocytoma, and the glioblastoma multiforme were the only diagnostic categories. The description of the glioblastoma multiforme as a pleomorphic cellular neoplasm with vascular proliferation, and usually necrosis, coincides with a definition widely applied today. In effect, the Ringertz system was adopted by the World Health Organization in its classification of 1979, although this may not be immediately evident upon reading its published description (ZÜLCH 1979). The glioblastoma is sequestered in a separate category of primitive and embryonal tumors since it was felt that the cell of origin of glioblastoma was neither always apparent nor, if apparent, always astrocytic. Nevertheless, these fibrillary neoplasms were categorized as either astrocytoma (WHO grade II), anaplastic astrocytoma (WHO grade III), or glioblastoma multiforme (WHO grade IV). The pilocytic astrocytomas of the cerebellum, hypothalamus, and optic nerve are assigned a grade I in light of their distinctive histologic appearance and relatively benign biologic behavior.

The utility of a three-tiered system cannot be denied since significant differences in survival between the three groups have been found since the time of Ringertz. A three-tiered system is utilized by at least two major cooperative groups and both have clearly defined differences in survival between the anaplastic astrocytoma and the glioblastoma multiforme (NELSON et al. 1983; BURGER et al. 1985). Both groups use necrosis as an important variable to separate these two grades. Most of these studies have been done on large tissue specimens obtained by open biopsy where the chances of finding necrosis and other histologic variables were maximized. It has been our experience subsequently that many tissue samples obtained stereotactically have overt features of anaplasia, and are undoubtedly glioblastomas, but lack necrosis or vascular proliferation. This underscores the problems of defining histologic grades in the face of limited samples, and the limitations of an inflexible dependence on single histologic variables for tumor grading.

A grading system used by the European Oncology Research Tumor Group is formulated along the lines of the WHO system, but recognizes four grades (BRUCHER et al. 1987). In this classification, the topographic distribution of well differentiated and poorly differentiated cells are considered.

The grade I neoplasm is a well differentiated lesion which is focal, whereas the grade II neoplasm is diffuse. The grade III tumor is an area of anaplastic astrocytoma in which foci of malignancy are noted. The glioblastoma multiforme is an anaplastic lesion in which only anaplastic cells are seen. The survival curves in patients with the grade III lesion of this system are significantly shorter than for grade III neoplasms of other grading systems. This is not surprising since these systems do not consider topographic features or the ratio of poorly to well differentiated cells. They would undoubtedly classify some of these grade III lesions as *glioblastoma multiformes*. This illustrates the difficulty in comparing grading systems and assuming that a grade III neoplasm in one system is equivalent to a grade III in another. This problem is especially acute due to the present concurrent usage of the Kernohan and three-tiered systems. In the former a grade III neoplasm is a glioblastoma multiforme, in the latter it is an *anaplastic* or *malignant* astrocytoma.

An interesting recent classification employs a system based in part on the number of specific histologic variables present, and also addresses the problem of grading in the face of the limited tissue obtained by needle biopsy (COHADON et al. 1985). By this system, necrosis is given a special prominence, whereas other variables are considered when present but are not required factors if the tissue is obtained by needle biopsy. As discussed above, it is of special interest that in this study, multivariate analysis considering the prognostic significance of duration of neurologic symptoms, age, and histologic grade concluded that the histologic grade did not contain independent prognostic significance.

Another recent study also computes a histologic grade on the number of variables present, but does so without any *weighting* of variables. (DAUMAS-DUPORT et al. 1989) By this classification the pathologist seeks the presence of cytologic atypia, vascular proliferation, mitoses, and necrosis. No qualitative information is demanded, only presence or absence. A neoplasm with no variables present is a grade I whereas a tumor with three or more is a IV. When the survivals of the patients of the groups are charted, four distinct and significantly different curves appear. The plots for the grades III and IV conform very closely to the survival curves of the anaplastic astrocytoma and glioblastoma multiforme of the Brain Tumor Cooperative Study Group. In principle this system is fundamentally different from most other grading systems in which certain variables such as necrosis are given special weight. In practice the variables are linked in such a way that one is unlikely to be forced into a diagnosis of an astrocytoma grade I when only necrosis is seen, since necrosis or vascular proliferation is almost always associated with cytologic atypia and mitoses.

This system would appear to have a number of advantages since it is as objective as possible and it identifies a group of extremely well differentiated lesions which are not given a separate category in the three-tiered system. Its applicability to tissue from other institutions and to small specimens obtained by needle biopsy remains to be seen, however. It has been our experience that mitoses may be extremely difficult to find in lesions which are, by all other criteria, overtly malignant. In addition, one would predict a reluctance of many pathologists to grade an obviously highly malignant neoplasm a grade III only because of the absence of one histologic variable, particularly if the specimen is obtained by needle biopsy. Thus, the advantageous objectivity of the system on the one hand makes it difficult for pathologists to add a subjective input into tumor grading on the other. This may be a disadvantage or advantage depending on one's viewpoint.

Oligodendrogliomas

Properties of Oligodendrogliomas and Limitations of Grading: The general properties and limitations of grading as applied to the fibrillary astrocytic neoplasms also apply to oligodendrogliomas. Thus, there can be considerable regional heterogeneity of these lesions wherein nodules or lobules of more cellular and more mitotically active neoplasms are scattered in the background of a better differentiated tumor. In addition, there is a well recorded tendency for the oligodendroglial neoplasms to progress to histologically more malignant lesions (MÜLLER et al. 1977). As in the astrocytic neoplasms there has been an inverse relationship between the age of the patient and the length of survival, although this relationship appears to be somewhat more loose than in the astrocytic neoplasms (BURGER et al. 1987a; KROS et al. 1988).

Review of Grading Systems of Oligodendrogliomas: Although unequivocal evidence that the astrocytic neoplasms can be usefully graded has been recognized for many years, it is only recently that similar sentiments have been expressed concerning the oligodendrogliomas.

SMITH et al. (1983) and LUDWIG et al. (1986) utilized a four-tiered grading system as applied to 323 cases from the Armed Forces Institute of Pathology. The lowest grade, grade A, was assigned if endothelial proliferation, necrosis, low nuclear cytoplasmic ratio, low cellularity, and minimal or absent pleomorphism were noted. Grade B was utilized if pleomorphism was present and/or if cell density and nuclear cytoplasmic ratio were high. Grade C neoplasms demonstrated pleomorphism, vascular proliferation, high nuclear-cytoplasmic ratios and high cell density. Grade D was given when all five features were present and high. The study demonstrated a significant relationship between the grade and survival, although there appeared to be little difference in survivals between grades B and C. Of the individual histologic features studied, only one factor, pleomorphism, survived as independently prognostic following multivariate analysis. KROS et al. (1988) have recently studied the *Smith* system and concluded that it is potentially useful and detects three groups for purposes of prognosis.

A study from the Cancer Registry of Norway (MØRK et al. 1986; LINDEGAARD et al. 1987) avoided the issue of grading and focused on the relation between specific histologic features and survival. In this study, the presence of necrosis and high cell density were associated with shorter survival whereas microcysts portended a longer clinical course. A shorter investigation of our own complemented the Norwegian Study and identified the number of mitoses, the presence of necrosis, nuclear cytologic atypia, and vascular proliferation as all having statistically prognostic significance when studied by univariate analysis. By the statistical manipulations of multivariate analysis, necrosis survived as the only independently prognostic factor.

When all these studies covering oligodendrogliomas are combined it is apparent that histologic prognostic factors include pleomorphism, necrosis, vascular proliferation, mitoses, cellularity, microcysts, and cytologic atypia. How these should be assembled into a grading system remains to be seen. As indicated above, the Smith system establishes three grades much as was done for the astrocytic neoplasms by Ringertz and his followers. In grading of the oligodendrogliomas the relationship of the oligodendroglioma to the glioblastoma multiforme becomes an important issue. Is the highest grade of the oligodendrogliomas comparable to the glioblastoma multiforme? In the referenced studies, care was taken to include only lesions with distinctive oligodendroglial features and it was our feeling that the lesions in our study were malignant oligodendrogliomas and distinct from the glioblastoma multiforme. Some support for this can be found in the survival curves of such patients with lesions having a survival somewhat longer than that for the typical glioblastoma. There is little question, however, that some oligodendrogliomas complete an evolution to an unequivocal glioblastoma multiforme with extremely high cellularity, undifferentiated cells, necrosis with pseudopalisading, and vascular proliferation. It has been suggested that this is, in part, accomplished by overgrowth of an astrocytic component. The more malignant oligodendrogliomas frequently acquire cells with conspicuous pink, strongly GFAP-positive cytoplasm. Characteristically these cells have the round nucleus and chromatin configuration of an oligodendrocyte and have scant processes. In some lesions, however, cells with more classic astrocytic features appear to arise and may be related to the progression of the neoplasm to the glioblastoma.

Newer Method for Grading

In light of the difficulties in grading systems as discussed above, it is appropriate that more objective means of tumor grading are being developed. The determination of a thymidine index provides such an objective means by detecting cells which are in the DNA synthesis or S-phase. This method can be done by injection of an isotope prior to surgery or by immersing small fragments of excised tissue into a solution containing the label (BROGGI et al. 1988). The former has the disadvantage that long exposure times are necessary because of the limited amount of isotope that can be administered to patients. The *in vitro* method circumvents this limitation, but requires small samples and amplifies the already difficult problem of sampling.

It is fortunate that immunohistochemical methods to identify cycling or proliferating cells have appeared. The technique which has received the most study is that using bromodeoxyuridine (BrdU). When injected pre- or intraoperatively, the monoclonal antibody detects this halogenated pyrimidine as incorporated into the DNA during the synthesis (S) phase of the cell cycle (HOSHINO et al. 1986). The proliferating cells are identified by the presence of the reaction product over the nuclei. The technique has been applied to a wide variety of gliomas and there has generally been a good correlation between the traditional grade and the staining index. When a high index is discovered in a lesion thought histologically to be well differentiated, suspicion is aroused that the lesion has been undergraded by the traditional means. It seems likely that this method will find special utility in such settings rather than in subclassifying overtly malignant neoplasms such as the glioblastoma. The BrdU technique has the distinct advantage that it can be used in paraffin-embedded tissue although it cannot be applied to archival tissue. It can be performed also *in vitro* by immersing small fragments of the excised neoplasm prior to utilization of the monoclonal antibody.

Another immunologic technique using the monoclonal antibody Ki-67 has complementary advantages and disadvantages to the BrdU method (BURGER et al. 1986). This nuclear antigen is identified by an immunohistochemical method and is expressed in all phases of the cell cycle except G0. The technique has the advantage that no injection of a compound is required. However, unlike the BrdU method, it cannot, at present, be performed on paraffin-embedded tissue. The experience with this technique is more limited than that of the BrdU, but we are optimistic that it will be a prognostically useful means for grading of gliomas (BURGER et al. 1986). A recent study of a series of neoplasms and the normal colonic epithelium compared the staining indices of BrdU and Ki-67 and found, as expected, a higher percentage of labeled cells from the Ki-67 method (SASAKI et al. 1988).

It seems inescapable that molecular approaches to the understanding of neoplasia will find applications in tumor grading. In the glioblastomas, the presence in approximately half of the lesions of amplification for the erb B oncogene (epidermal growth factor receptor) have been discovered (BIGNER SH et al. 1988e). A clinicopathologic study has suggested that there is no differences in survival between patients with glioblastomas with or without amplification. The study also noted the general paucity of amplified cases within the oligodendroglial or anaplastic astrocytoma groups, but in the latter category there were cases with amplification. These may well represent undergraded glioblastomas which, because of tissue sampling, did not contain the classic histologic features of the glioblastoma. Thus, the detection of gene amplification in malignant gliomas may prove to be a prognostically useful means for grading of gliomas. It seems reasonable that ultimately the prognosis can be estimated by attention to molecular observations although their variability and instability may require a grading system of their own.

CHAPTER XIV

Grading of Brain Tumors According to the WHO Classification: Applicability to Clinical Medicine

JEAN-MARIE BRUCHER, *Department of Neuropathology, University of Louvain, Brussels, Belgium*

Grading Systems

A grading system of tumors is a response to the daily demand of clinicians, who need accurate informations about the probable evolution and prognosis of each operated tumor. The *four-grade system* introduced by BRODERS (1926) for the classification of carcinomas and applied by KERNOHAN et al. (1949) to the classification of gliomas has received a wide acceptance because of its simplified character and its biologic meaning. Nevertheless, in spite of some further improvements (KERNOHAN and SAYRE 1952, SAYRE 1956), this grading system remained unclear because it was based on rather complex and difficult histologic criteria (number of mitotic figures, percentage of dedifferentiated tumor cells, extent of necrosis and vascular proliferation, and degree of pleomorphism). Moreover, many gliomas with local variations of these histologic characteristics were frequently given mixed grades (grade I-II or grade III-IV or even grade II-IV) by many pathologists.

The *four-grade system* developed by SMITH et al. (1983) for oligodendrogliomas was based on five histologic criteria among which the mitotic rate was not included. Moreover, the survival curves showed no significant difference between grades B and C.

The *three-grade system* developed by RINGERTZ (1950), and widely used, namely by some American oncology groups (NELSON et al. 1983) and the National Brain Tumor Study Group (BURGER et al. 1985), categorizes astrocytic neoplasms as astrocytoma, anaplastic astrocytoma and glioblastoma, corresponding to astrocytomas grade I, grade II and grade III-IV respectively. The histologic criteria (hyper-cellularity, pleomorphism, vascular proliferation, and necrosis used by the National Brain Tumor Study Group are ill-defined and difficult to apply. For instance, the key criterion used to distinguish glioblastoma from anaplastic astrocytoma was the presence of necrosis, but BURGER et al. (1985) admitted that small foci of necrosis in some cases of anaplastic astrocytoma "did not appear to correlate with poorer survival". In a large study of 184 cases of malignant gliomas considered as glioblastoma multiforme, BURGER and VOLLMER (1980) already admitted that necrosis, endothelial proliferation and cellularity "did not have any consistent relationship with prognosis". Moreover, they showed that pleomorphism is not a good histologic criterion since "giant cell neoplasms were associated with prolonged survival". In spite of the limited value of these "major histologic criteria", BURGER et al. (1985) could conclude that "the subclassification of malignant gliomas into the anaplastic astrocytoma and the glioblastoma multiforme defines groups of patients that are significantly different in regard to age, duration of symptoms, and length of survival".

In these grading systems, the *histologic grades* were considered as corresponding to biologic *stages* or *steps* during the natural evolution of the neoplastic proliferation, and the glioblastoma was classified as a highly malignant astrocytic tumor. In fact the glioblastoma contains not only variable amounts of GFAP-positive cells, but also numerous undifferentiated glial cells, the origin of which is difficult to assess.

In the WHO classification (ZÜLCH 1979), the tumor entities were simply described without theoretical considerations of histogenesis, except for anaplastic subtypes which were systematically recognized for most of the tumor types. Each tumor entity received a *biologic grade* (from I to IV) and grade III was given to the anaplastic subtypes. This *biologic grading system*, which had been developed by ZÜLCH and WECHSLER (1968) and by ZÜLCH (1975), was not described again in the *Blue Book* of the WHO classification, but the definition of this grading was recalled one year later by ZÜLCH (1980). The meaning of this *biologic grading* is only to give information about the prognosis. As also emphasized by SMITH et al. (1983), "the idea of attaching a prognostic estimator to the tumor's name has merit".

Thus, in the WHO classification of brain tumors (Fig.1), the grades are not considered as successive stages in the natural evolution of the tumors. The prognostic estimation is based on the present growing potentialities of the tumor considered as a whole, including both quiescent and active parts of the neoplastic process. One has to remember that the growth activity depends not only on the responsible agent but also on the defense mechanisms of the host. The mixture of quiescent and active parts in a tumor is reflecting an intermediary degree of malignancy, which is given grade III in contrast to grade IV tumors, which are fully malignant in all their parts and from the beginning of the neoplastic process. This is the reason why the glioblastoma often shows rather sharp limits and may be confused with a metastatic lesion.

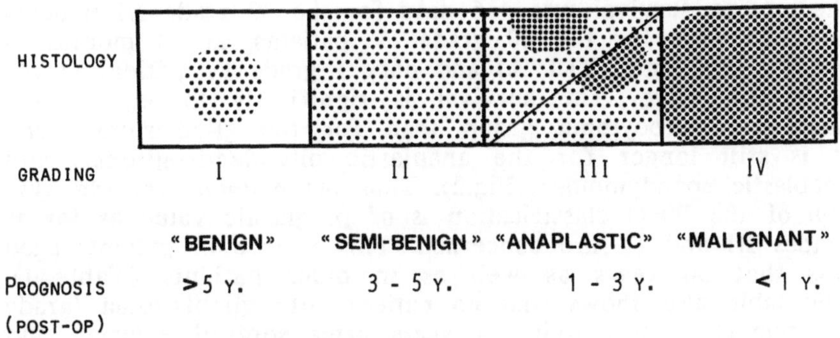

Figure 1 Biological grading in neuro-oncology

On the other hand, well delineated or even encapsulated, benign tumors are biologically different from slowly growing tumors with diffuse or multifocal spread such as the astrocytoma and the oligodendroglioma. The first group is graded I, which means that these tumors can be surgically cured if they are operable. The second group is graded II since the neoplastic process will probably continue its progression outside the operated area.

Prospective Analysis

In my opinion, the *biologic grading* and the systematic recognition of anaplastic subtypes are the most useful points of the WHO classification of tumors of the central nervous system. This clear distinction between grade III and grade IV glial tumors, i.e. between anaplastic gliomas and glioblastoma, had to be checked by studying a large series of brain tumors. This opportunity was recently offered by the Brain Tumor Group of the European Organization for Research and Treatment of Cancer (EORTC). With the help of O. Dalesio (statistician) and G. Solbu (data manager), we were able to perform a prospective analysis of 202 patients with grade III anaplastic gliomas or grade IV glioblastomas. The surgical excision was followed by radiation therapy (6000 rads over a period of 6 weeks); 103 patients received adjuvant chemotherapy (VM-26 + CCNU: 3 courses in 12 weeks) before the radiation therapy. The results in both treatment arms showed no significant difference.

The neuropathologic material from these 202 brain tumors was reviewed and classified according to the above mentioned criteria of the WHO classification and without knowledge of the survival. The curves showing the *survival from surgery* demonstrate a clear difference between all the grade III and IV tumors including cases both with and without adjuvant chemotherapy. In the series without adjuvant chemotherapy, the median survival was 416 days for the 39 grade III patients and 305 days for the 60 grade IV patients; at 24 months, 9 grade III patients (23.1 %) and only 1 grade IV patient (1.7%) were still alive. The survival is significatly longer for the anaplastic astrocytoma group than for the glioblastomas and it is still longer for the anaplastic oligodendrogliomas and anaplastic ependymomas (Fig.2). Thus the *biologic grading system* of the WHO classification is of prognostic value as far as grades III and IV are concerned. This is true in patients aged less that 50 years as well as in older patients (Table I). The table also shows that no patient with glioblastoma (grade IV tumor) is still living 3 years after surgical excision and radiation therapy, with or without chemotherapy.

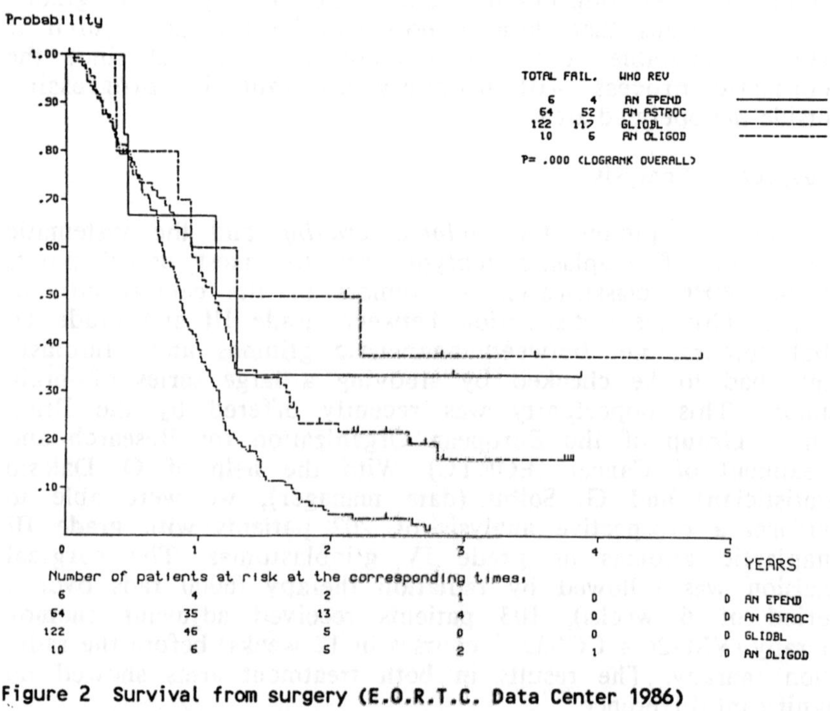

Figure 2 Survival from surgery (E.O.R.T.C. Data Center 1986)

TABLE I Survival from surgery: 201 cases

Age at surgery (years)	<50		≥50		16 - 75	
	Grade 3	Grade 4	Grade 3	Grade 4	Grade 3	Grade 4
Number of patients	34	35	46	86	80	121
Survival: 2 years	15 (44.1%)	3 (8.6%)	5 (10.9%)	2 (2.3%)	20 (25.0%)	5 (4.1%)
3 years	6 (17.6%)	0 (0.0%)	1 (2.2%)	0 (0.0%)	7 (8.7%)	0 (0.0%)

The data published on a larger series by BURGER et al. (1985) also show highly significant differences in survival between patients with glioblastoma (1265 patients) and those with anaplastic astrocytoma (175 patients). Nevertheless the comparison of these data with our results reveals some differences :

1) The 2-year survival of patients with anaplastic astrocytoma ranges between 38 and 50 % in the study of BURGER et al. (1985) and is only 20% (13 out of 64 cases) in our material.
2) The 2-year survival of patients with glioblastoma ranges between 8 and 12 % in the Burger study and is only 4% (5 out of 122 cases) in our series.
3) The ratio between the number of patients with glioblastoma and the number with anaplastic astrocytoma is 1265/165 = 7.2 in the Burger study material and 122/64 = 1.9 in our material.

These differences can be explained by the fact that the *definitions* of glioblastoma and anaplastic astrocytoma are not exactly similar in both studies. Astrocytomas with moderate hypercellularity and moderate pleomorphism are considered as anaplastic astrocytomas by Burger, but they would probably be classified as astrocytomas by us if signs of focal anaplasia (with hypercellularity, mitotic figures and less differentiated cells) are not found. Thus the prognosis in our series of anaplastic astrocytomas will be less favourable if such pleomorphic astrocytomas are not included. On the other hand, cases with foci of marked hypercellularity, marked pleomorphism and even necroses will still be considered by us as anaplastic astrocytoma if quiescent areas outside of these foci demonstrate a lower malignancy of the whole tumor and a better resistance of the host. If these cases were incorporated in the glioblastoma group its survival curve would be improved.

We do agree with the conclusion of BURGER et al. (1985) that: "The distinction between the anaplastic astrocytoma and the glioblastoma multiforme seems legitimate and important for the performance of large randomized clinical trials". Nevertheless, in spite of accurate statistical analysis, the results are not entirely comparable from one study to the other because the definition of such common glial tumors as anaplastic astrocytoma and glioblastoma remains to be improved.

Histologic Criteria

The WHO classification should include a better definition of the histologic criteria for glial tumors. Even the *concept of glioblastoma* needs a consensus. According to ZÜLCH (1986), it is a tumor "defined as having a primary origin and growth" and thus it is not the end stage of a progressive dedifferentiation process in an astrocytoma as suggested by the grading system of KERNOHAN et al. (1949). BURGER et al. (1985) also conceded that "as clinical entities, the anaplastic astrocytoma and glioblastoma may be more individual than is often assumed". In a large series of 129 necropsies of patients with a previous histologic diagnosis of cerebral glioblastoma, RUBINSTEIN (see RUSSELL and RUBINSTEIN 1988) could reclassify 78 cases as *pure* or *primary* glioblastomas in the absence of any definite evidence of a pre-existing diffuse astrocytoma. The presence of numerous, GFAP-positive, astrocytic cells in many glioblastomas cannot be considered as a proof of the astrocytic origin of the neoplastic cells since focal differentiation may occur in malignant tumors. Moreover, the tumor cells of the glioblastoma often are "so undifferentiated that their astrocytic origin may hardly be recognized in sections" (RUBINSTEIN 1972b).

On the other hand, the histologic criteria of the *anaplastic transformation* should be discussed. From a logical point of view, the most reliable criterion of anaplasia is the combination of high cellularity, poor differentiation and mitotic figures. Other frequently used criteria, such as pleomorphism, hyperchromatic nuclei, necrotic foci, endothelial proliferation and infiltrating character, are inconstant and unreliable, at least as isolated characteristics. For instance, in a glioblastoma, large pleomorphic glial cells, which are commonly considered as signs of anaplasia, are often rich in GFA protein and thus more differentiated than the small cells. These large, polymorphic cells are considered by ZÜLCH (1956) as regressive cell changes. Endothelial proliferation, often related to necrosis, can be encountered in non-malignant gliomas such as the cystic pilocytic astrocytoma.

SCHERER (1940) has emphasized the frequent occurrence of anaplasia in astrocytomas. With the modern imaging techniques (CT scan, NMR) it is often possible to detect both components of anaplastic gliomas: the diffuse, quiescent parts and the anaplastic foci. The problem of the histologic criteria for the anaplastic transformation thus becomes of major importance, particularly when stereotactic biopsies are performed in several parts of the tumor.

Discussion of Chapter XIV

DR. KIRKPATRICK: I would like to just confirm with you, Dr. Brucher, whether or not you said that an anaplastic focus within a lesser grade tumor might bias the grading of glioblastoma and make it appear to offer a better survival. Is that your opinion?

DR. BRUCHER: Yes.

DR. KIRKPATRICK: I have been taught to grade by the worst area I can possibly find. Perhaps we can discuss that later.

DR.KEPES: You said hypercellularity is the most important feature. Do you mean only in astrocytomas or in other gliomas too?

DR. BRUCHER: No. I showed that feature in oligodendrogliomas and in the ependymomas as well.

DR. KEPES: Oligodendrogliomas, even the low grade ones, can be quite cellular.

DR. WECHSLER: I would like to make only some brief complementary statements. One of the major concepts developed during the past decade in neuro-oncology was the introduction of the oncogene theory and the concept that malignant gliomas in human patients should be considered a brain tumor group with the very striking feature of cellular heterogeneity. This has been documented by a number of scientists, including Dr. Bigner's group at Duke. Heterogeneity has to be conceived at different levels: at the tissue level, the cellular level, the immunologic level, the biochemical level and the genetic level.

As far as the application of immunohistochemistry for tumor grading is concerned, There is no doubt that this is a very difficult field of research. We believe, however, that immunohistochemistry for certain antigens, including the MMA (Leu M1) epitope, protein kinase C (PKC) and epidermal growth factor receptor, may be of importance here, especially when adjacent sections are studied simultaneously for proliferation-associated antigens (see REIFENBERGER et al. 1987). Disregarding the undisputed value of intermediate filaments immunohistochemistry for differential diagnostic purposes, the expression of GFAP and vimentin in human gliomas is of limited value for tumor grading. However, some authors have reported a tendency towards increased expression of vimentin and decreased levels of GFAP immunoreactivity in high-grade astrocytomas compared with low-grade ones. Nevertheless, if one studies GFAP and vimentin expression in frozen sections of human astrocytomas, it has been our experience that this tendency is fairly vague and that most astrocytomas show co-expression of GFAP and vimentin irrespective of their grade of malignancy. May I therefore offer some more remarks with respect to the proliferation index in human gliomas? The first publication on this subject by BURGER, SHIBATA and KLEIHUES (1986) calculated the percentage of proliferating cells in 40 neoplasms of the nervous system using the monoclonal antibody Ki-67. The labelling index for anaplastic astrocytomas and glioblastomas ranged from 1.6-12.4% with an average of 6.2% for nine glioblastomas. GIANGASPERO et al. (1987) studied six glioblastomas with a mean proliferation index of 28% (standard deviation 9.8, range 10-40%).

A summary of our data is illustrated in Table I of Chapter IX. If you calculate the percentage of labelled cells in different areas of the tumor tissue, the proliferation index (PI) for grade I pilocytic astrocytomas is less than 1% with a range and standard deviation below 1. Grade II astrocytomas had a PI of 1.4% with a range of 0.8% and a standard deviation of 2.4. For anaplastic astrocytomas of grade III a mean proliferation index of approximately 9% with a fairly large standard deviation was found. If one compares astrocytomas and oligodendrogliomas of WHO grade II, the oligodendrogliomas had a PI of 4.2%, which is thus higher than that of the astrocytomas (1.4%).

It is of interest that primary and recurrent oligodendrogliomas with anaplasia representing grade III gliomas showed a dramatic increase in the proliferation index with a range of 10-20%. Therefore, it was of some surprise that 18 glioblastomas, i.e. grade IV gliomas, presented a fairly large range of 0-22% with a mean proliferation index of only 7.0%. This indicates that proliferative heterogeneity is a striking feature in the glioblastoma group.

Finally, I would like to demonstrate the neuropathologic features of two patients with recurrences. The first interesting case is a late fatality of a primary isomorphous oligodendroglioma of grade II in a 70 year old female patient. There had been 24 years between the operation of the primary tumor and the recurrence. The recurrent tumor maintained all the essential features of an oligodendroglioma including calcification and was composed of areas with a low proliferation index and areas of strikingly increased proliferation (mean 10%, range 3-15%).

The second case history is even more interesting. It is that of a 46 year old female with a recurrent astrocytoma. The primary tumor occurring in the left frontal lobe at the age of 41 was a typical gemistocytic astrocytoma without any signs of anaplasia (WHO grade II). Four years after resection this neoplasm recurred demonstrating a dramatic increase in anaplasia. There were only residual areas reminiscent of the primary isomorphic astrocytoma. The major part of the tumor now consisted of densely-packed, polymitotic, small anaplastic glioma cells. Numerous foci of necrosis were discernible. The patient refused postoperative radiotherapy and then underwent adjuvant intra-arterial ACNU chemotherapy. However, the second recurrence occurred one year later and was again operated on. At this time histopathologic examination revealed a highly anaplastic glioma which was graded as corresponding to WHO grade IV. This assessment was underlined by the determination of the Ki-67 labeling index which exhibited the exceedingly high value of 53%. The patient died one month later and necropsy revealed not only residual tumor tissue at the primary site but also extensive ependymal and meningeal blastomatosis as well as large solid metastatic lesions in the cerebellopontine angle and the falx cerebri. This case, which clearly illustrates the transformation of a semi-benign astrocytoma into a highly anaplastic glioma with fatal outcome, also demonstrates that there are cases of malignant glioma not belonging to the glioblastoma group which exhibit the behavior of WHO grade IV tumors including extremely high proliferative activity as indicated by Ki-67 staining. This possibility is, however, not considered in the WHO grading system which ascribes the grade IV only to gliomas belonging to the glioblastoma group.

Let me summarize our present approach to tumor classification of human brain tumors in clinical neuropathology. For a number of reasons we are very much interested in studying brain tumors both in unfixed frozen sections and formalin-fixed paraffin material. In addition to the conventional histopathologic examination this offers the opportunity for study, in serial frozen sections, of the differentiation markers such as intermediate filaments or desmoplakins as well as other antigens such as protein kinase C, epidermal growth factor receptor and markers relevant to the detection of proliferating cells, i.e. the monoclonal antibody Ki-67, which reacts with a nuclear protein expressed in G_1-, G_2-, S- and M-phases of the cell cycle. Of course, we are aware that these data are still preliminary and that future clinicopathologic follow-up studies must show whether immunohistochemically determined proliferation indices correlate with survival time. Hopefully, further and better quantitative methods for tumor grading will become available in the future.

CHAPTER XV

Clinical Correlations of Malignant Ependymomas

LUCIEN J. RUBINSTEIN, *University of Virginia School of Medicine, Charlottesville, Virginia.*

The points of the following presentation are two-fold. The first illustrates the need to separate the embryonal central neuroepithelial tumors from those which are essentially composed of more differentiated, or adult-type cells in which anaplastic change has occurred. This is because those two distinct categories of neoplasm may display different biological, and therefore clinical, behavior. The second point is that, whereas clear clinicohistological correlations have been established in the astrocytomas, no such concordance seems to be apparent in the case of ependymomas that show histologic features of malignancy, provided the ependymoblastomas are excluded from the analysis.

It is widely believed that an important determinant in the clinical behavior and prognosis of patients who harbor an ependymoma is the histologic grade of malignancy of the tumor and that an increase in histologic anaplasia corresponds to a higher grade of clinical malignancy (MABON et al. 1949; KRICHEFF et al. 1964a; KRICHEFF et al. 1964b; BARONE and ELVIDGE 1970; SALAZAR et al. 1975; DOHRMANN et al. 1976; LIU et al. 1976; MØRK et al. 1977; AFRA et al 1983; PIERRE-KAHN et al. 1983; SALAZAR et al. 1983; LEIBEL and SHELINE 1987; SHAW et al. 1987). The microscopic features of anaplasia comprise increased cellular density, cytologic pleomorphism, loss of distinctive differentiating structures, increased mitotic activity, spontaneous large or small necroses, and endothelial vascular proliferation. It is, however, apparent that, in the reports just cited, a number of exceptions and many inconsistencies in clinicopathologic correlations can be found and that there is a lack of agreement on the nomenclature to be applied to the less differentiated tumors. This is because, with one exception (LIU et al. 1976), no distinction seems to have been made between the ependymoblastomas and those otherwise well-differentiated ependymomas which show histologic evidence of anaplasia and for which the term *malignant ependymoma* appears to be more appropriate.

The *ependymoblastoma* has been defined as a highly cellular, embryonal form of ependymal tumor that occurs most often in infants and children under the age of five (RUBINSTEIN 1970; MØRK and RUBINSTEIN 1985. Unlike intracranial ependymomas, ependymoblastomas tend to show a preference for the supratentorial compartment (LIU 1976; MØRK and RUBINSTEIN 1985; CRUZ-SANCHEZ et al. 1988). Their distinctive feature is the presence of ependymal rosettes and tubules, which are often, though not invariably, numerous. The rosettes are usually composed of cells forming multiple layers, and mitotic figures are often demonstrable in a juxtaluminal position. In this respect they resemble the Flexner-Wintersteiner rosettes of retinoblastoma (WILLIS 1967) and differ from those found in ependymomas, in which the rosettes are surrounded by only one, or only occasionally two layers of cells and are generally devoid of mitotic figures. These rosettes have been termed *ependymoblastic* (VANDENBERG et al. 1975). They are similar to those found in the early differentiating neuroepithelial elements of the experimental transplantable mouse testicular teratoma, in which they are transitional between the primitive medullary rosettes and the mature ependymal rosettes (HERMAN et al. 1975). The electron microscopic features of a carefully studied and typical example in man (LANGFORD 1986) were identical to those found in the zones of ependymoblastic differentiation in the transplantable mouse teratoma.

In addition, ependymoblastomas may include primitive polar glial cells (RUBINSTEIN 1970), and a very small number of these cells may occasionally be positive for glial fibrillary acidic protein (CRUZ-SANCHEZ et al 1988). Immunopositivity of the tumor cells for vimentin and for the S-100 protein has also been recorded (CRUZ-SANCHEZ et al. 1988).

Biologically, the ependymoblastoma differs from the classic ependymoma by the frequency with which focal microscopic invasion of the adjacent brain and leptomeningeal involvement are demonstrable. Widespread metastases may occur in the cerebrospinal subarachnoid space (MØRK and RUBINSTEIN 1985; WADA et al. 1986). Thus it behaves in the same manner as a cerebellar medulloblastoma. Examples have been recorded in which extraneural metastases developed (GLASAUER and YUAN 1963; CAMPBELL et al 1984; MØRK and RUBINSTEIN 1985). Although these tumors are rare, there is general agreement that the clinical prognosis is very poor. The median postoperative survival times range from 12 months (MØRK and RUBINSTEIN 1985) to 20 months (LIU et al. 1976). A 100 percent mortality has been recorded within 3 years of diagnosis (LIU et al. 1976).

In the present study we have therefore specifically excluded from the analysis the group of ependymoblastomas, whose personally examined material we have previously reviewed (MØRK and RUBINSTEIN 1985). We have attempted to correlate 15 personally studied examples of other histologically malignant ependymomas with the development of local tumor recurrence and with the postoperative patient survival times. Our findings suggest that, in this particular category, the histologic features usually regarded as predictive of clinical malignancy are of no value in helping to anticipate subsequent tumor recurrence or other postoperative clinical behavior.

The histologic and clinical details on these 15 cases are currently being reported elsewhere (ROSS and RUBINSTEIN 1989). Briefly, all but one of these cases were referred to us from 1967 through 1986, specifically because of the presence of malignant histologic features. The patients ranged in age from 1 to 27 years, the median age being 11 years. There was no significant difference in age incidence between patients with a supratentorial tumor and those with an infratentorial tumor. The posterior fossa was the most frequently involved location (10 cases). Eight intracranial tumors were related to one of the ventricles; two were extraventricular and in four cases the relationship to a ventricular cavity was uncertain.

All the tumors included areas of typical ependymoma, but in addition presented fields with variable degrees of cellular atypia and nuclear pleomorphism. Where the typical ependymomatous pattern was lacking, the tumors grew in homogeneous, densely cellular sheets. Almost all the examples contained areas of necrosis, and almost all displayed vascular endothelial proliferation, which was often marked. Mitotic figures were easily found in all cases, and were particularly numerous in the more cellular areas.

All the patients had had surgical excision of their growth, either subtotal or complete. Postoperative radiotherapy was administered to 12 of the 15 patients, and chemotherapy to five. Details on the modalities of treatment varied greatly from case to case.

The results of this survey indicated that 10 (i.e. two-thirds) of the patients were alive from 15 months to 14 years after the original operation (median postoperative survival time: 8.8 years). Seven of the patients were alive after 6 years or longer. Most of them were disease-free, but one patient is blind, though otherwise well. In one patient, recurrence took place 11 years after operation, and the recurrence this time was histologically benign. One patient developed a malignant meningioma, which was presumably induced by postoperative radiation, 14 years after the removal of a malignant supratentorial ependymoma.

Five of the patients, however, (i.e. one-third in this series) died of their tumor, all of them with documented local clinical and radiologic recurrence. Their postoperative survival times ranged from 13 months to 6 years, with a median survival time of 2.5 years. Metastastic development, either within or outside the central neuraxis, is not known to have occurred in any of the patients in this series.

The prognostic and therapeutic implications to be drawn from these flndings are significant. They emphasize the need for exact histologic identification. The diagnosis of ependymoblastoma requires treatment as for a medulloblastoma, including the administration of whole central neuraxis radiation, and possibly chemotherapy. Malignant ependymomas, on the other hand, carry a much less predictable prognosis: the latter is often favorable, provided surgical excision has been successful, but fatal recurrence is to be anticipated in about one-third of the cases. Most of the patients in this study were given postoperative radiotherapy, but although it has been stated that this modality substantially improves the outcome of patients with intracranial ependymomas in general (LEIBEL and SHELINE 1987), there are no controlled data that exactly determine its efficacy.

The highly variable prognosis of malignant ependymomas, and the absence of correlation between histologic features, site of the growth and likelihood of tumor recurrence, contrast with the clear-cut correlations that have been established in cerebral astrocytic gliomas((NELSON JS et al. 1983; BURGER et al 1985; FULLING and GARCIA 1985; NELSON DF et al. 1985). It is noteworthy that similar discrepancies between histologic picture and clinical behavior have long been known in respect to some of the microscopic features of oligodendrogliomas(ROBERTSandGERMAN1966;WEIRandELVIDGE1968;MØRK et al 1986), and that they have more recently been recorded for some of the giant cell subependymal astrocytomas associted with tuberous sclerosis (CHOW et al. 1988) and confirmed again in the case of the rare astroblastoma (RUSSELL and RUBINSTEIN 1988).

General Discussion of Part Two

DR. KIRKPATRICK: I would like to open the discussion by raising some questions myself. The first question I thought about as I was auditing these very exciting talks was, "How are we to structure the classification so as to incorporate the new molecular biology?" The caveat here, I think, is that in the past when new developments have come along and the terminology has been changed in order to incorporate the new developments, then the terminology inevitably became dated itself within a relatively short period of time. My bias would be to attempt to use the classical terminology but at least consider how we are to incorporate the new molecular biology into it, or perhaps alongside it.

Another thing we need to consider is where to put glioblastoma multiforme. In my opinion, it is a mistake to put it separately, as has been done in the *Blue Book*. Dr. Bigner referred to it as a final common pathway, and several speakers have emphasized the progression of oligodendrogliomas to glioblastoma. My bias is that it is primarily an astrocytic tumor, so I hope that will be considered appropriate for that group.

With respect to grading, it seems to me that insufficient attention so far has been given to those very important intermediate grades of astrocytoma. We also need to consider the application of the grading system to the less frequent tumors, oligodendroglioma and ependymoblastoma. However, it may be that we are dealing with small numbers of cases which do not reach statistical significance.

We should consider very strongly the issue of whether any of these gliomas are benign. In my opinion, significant disservice has been achieved to tell insurance companies, governments, and other funding agencies that any of these gliomas are *benign.* I would like for this august body to seriously consider that it is very likely that all of them should be called malignant. I am thinking in particular of an unfortunate patient, who because he happens to have a fairly well differentiated ependymoma of the spinal cord, cannot qualify for insurance reimbursements.

This is the manner in which the deliberations that we have in this kind of a meeting do indeed have implications for the outside world. I was pleased to hear Dr. Davis make similar remarks.

The sampling problem, which we inevitably face with the improvements of surgical technique into stereotaxic modalities, and the increasing use of needle biopsies, needs to be considered. I think that we would want to encourage the neurosurgeons to use avenues which have less morbidity. At the same time, we must recognize that this may limit us to some extent. Dr. Kepes, for instance, criticized strongly the use of needle biopsy instead of an open procedure in which there is access to more tissue. However, we need to consider what effect the use of these small samples will have on the grading system.

We have also not yet really addressed the issue of the correlation of our grades with the scanning techniques which are being used so widely and so effectively. It is my impression that the low grade gliomas in particular are being identified by these imaging techniques and we are often confronted with small biopsies of material which may be very difficult to categorize. Finally, I think we will have to decide just where to include the neural tumors which are becoming increasingly recognized.

DR. ISHIDA: In the WHO classification, anaplastic astrocytoma is defined as "astrocytoma containing areas of anaplasia." Follow-up study of six cases diagnosed as such revealed that five cases had survived more than one year and seven months. RORKE et al. (1985) defined anaplastic astrocytoma in a different way: "A tumor formed by astrocytes demonstrating varying degrees of anaplasia. The tumor may contain foci of necrosis, but they are not of pseudopalisading type." This category seems broader than that of WHO. It may thus allow inclusion of an astrocytoma in which anaplastic cells are found throughout the tumor, an astrocytoma formed by anaplastic fusiform cells, an astrocytoma composed of predominantly uniform rounded anaplastic cells, and an astrocytoma with intracytoplasmic hyaline inclusions and occasionally with desmoplasia.

The follow-up study of 31 cases with these types of astrocytoma disclosed death in 12 cases within one and a half years following the first surgery and that three cases displayed typical histologic features of glioblastoma at autopsy. There will be a considerable overlap between anaplastic astrocytomas and glioblastomas if we use the term *anaplastic astrocytoma* in this broader fashion.

Intracytoplasmic hyaline inclusions found in the pleomorphic astrocytomas are, in electron micrographs, divided into two types, Rosenthal fiber type and residual body type. It is noteworthy that the pleomorphic astrocytoma with hyaline inclusions has a favorable postoperative course without signs of recurrence in a young subject.

DR. SZYMAS: Dr. Kleihues told us about the significance of GFAP positivity in astrocytic differentiation in medulloblastoma and so did Dr. Ishida. I believe we should speak about the glial differentiation and not only about the astrocytic one. In a couple of tumor series, we have obtained the following results (GOTTSCHALK and SZYMAS, 1987):

GFAP positivity

Astrocytomas (224 cases)	100%
Oligodendrogliomas (100 cases)	64%
Glioblastomas (112 cases)	94%
Ependymomas (98 cases)	72%
Medulloblastoma (114 cases)	20%
Monstrocellular sarcomas (18 cases)	91%
Pleimorphic Adenomas of the salivary gland (10 cases)	90%

So at the present time we cannot be sure that GFAP positivity always signifies astrocytic differentiation.

I recently published a paper about GFAP positivity and prognosis in medulloblastoma (SZYMAS et al. 1987b). I could not find any correlation either in the GFAP-positive or the GFAP-negative tumor series. BURGER and his coworkers (1987a) have also published a paper about GFAP positivity in oligodendrogliomas, and they could not see any correlation there either.

In some other work (MARTIN et al, 1986) we found good correlation between the GFAP immunoreactivity, the duration of anamnesis and the survival time after surgery in both anaplastic astrocytoma and glioblastoma.

I am embarrassed because in some other published work with REIFENBERGER et al. (1987), we were unable to find any correlation between GFAP positivity and the grade of malignancy in astrocytoma. There was equally strong positivity in astrocytomas grade I, II, as well as III. I think it is because of some small technical differences between these two papers. In the series done with Martin and Gottschalk the slides were treated with Triton X-100 before the incubation with antibodies. In the other series with Reifenberger, slides were digested with trypsin before the first incubation.

There was also discordance between the histochemical and biochemical studies of GFAP. In a short series of glioblastomas (SZYMAS, 1986) we obtained two fractions of GFAP. One is insoluble and this is designated as the *A* fraction. The other is water soluble and is designated as *B*. In some cases we found only the soluble fraction, while in three cases we got both.

Table I indicates a comparison of the contents of water soluble and water insoluble fractions from the literature and from my own cases. In the normal white matter (ENG and DeARMOND, 1982) the proportion between water soluble and water insoluble fraction is 20%. In the reactive white matter the proportion is 50% (ENG and DeARMOND, 1983), and in multiple sclerosis also 50% (ENG and DeARMOND, 1982). In pilocytic astrocytoma it is 61% and in anaplastic astrocytoma and glioblastoma this proportion rose to 90% and declined in normal white matter (SZYMAS et al, 1987b). Consequently, I am really not sure what we can learn about GFAP from immunohistologic study.

I used immunoelectron microscopy to search for the answer to this question (SZYMAS et al, 1987b). This study was made on paraformaldehyde fixed material. We found heavy GFAP positivity connected with glial filaments in the reactive astrocytes. You can also see good positivity of astrocyte glial filaments in fibrillary astrocytoma. However, in glioblastoma you can find positivity only in some processes and this positivity is connected only with glial filaments. So I can say that in material fixed with paraformaldehyde, made permeable for antibodies with Nonidet-40 (non-ionic detergent, the same class as Triton K-100), GFAP positivity is only connected with glial filaments. One can say also that in this kind of immunohistochemical preparation, GFAP positivity indicates maturity of the glial tumor.

TABLE I Proportion (%) of water-soluble to water-unsoluble GFAP

Material	GFAP water-soluble mg/g w.w.	GFAP water-unsoluble mg/g w.w.	Total mg/g w.w.	(%)
Normal white matter	0,254[a]	-	-	20%[b]
Reactive white matter[c]	2,06 + 0,08	1,98 + 0,9	4,04	48%
Multiple sclerosis[d]	-	-	-	50%
Astrocytoma, pilocytic[c]	6,19 + 0,17	3,98 + 0,19	10,17	61%
Astrocytoma, anaplastic[c]	16,50 + 0,58	6,41 + 0,31	22,91	72%
	5,19 + 0,13	2,00 + 0,07	7,19	72%
Glioblastoma	2,03 + 0,02	-	-	-
	3,01 + 0,18	-	-	-
	3,91 + 0,12	-	-	-
	4,84 + 0,21	1,80 + 0,04	6,94	73%
	18,30 + 0,75	2,98 + 0,12	21,28	86%
	2,17 + 0,14	0,25 + 0,02	2,42	90%

w.w. = wet weight
a - ENG LF, et al. (1976)
b - ENG LF, DeARMOND SJ (1982)
c - SYMAS J (1986)
d - ENG LF, DeARMOND SJ (1983)

DR. KEPES: I was a little surprised that hypercellularity was singled out as perhaps the most important piece of evidence for malignancy. I think for that one has to nail down quite firmly first what kind of tumor one is talking about because the cellularity of a grade I oligodendroglioma greatly exceeds that of a grade II astrocytoma. It is probably not so easy to tell them apart because astrocytomas can look very much like microcystic oligodendrogliomas. Therefore I see a certain danger in projecting the spotlight on cellularity alone.

I also wondered why Dr. Brucher placed ependymoma in the first group when he tried to classify brain tumors according to general behavior and gave them grades. For example, neurolemmoma got grade I, and the whole group of ependymomas somehow also got into I. Although Dr. Rubinstein gave us some rather encouraging data on certain ependymomas, I am not sure that in the scheme of I to IV this would place ependymomas in the same category as neurolemmomas.

So, I have these two questions first, and then since I am at the podium, I want also to make a remark about what Dr. Kirkpatrick said concerning needle biopsies. I do have a bias because I fear that it helped us to come full circle in about eight years in the history of brain tumors. In the following I will show you why I think that:

Historic Phases in Management of Gliomas

Phase I	The patient seems to have a glioma. Too bad... There is nothing that doctors can do. Let us pray...
Phase II	Brave neurosurgeons are removing parts of the glioma
Phase III	Resect the tumor and irradiate the remnants, regardless of the histologic type.
Phase IV	Different tumors must be treated with different types of chemotherapy and irradiation. For success we must have a diagnosis that is absolutely exact!
Phase V	For this you must give us a correct diagnosis from a 2 x 1 mm stereotactic needle biopsy!
Phase VI	(In the United States) or else you may be sued for 4 million dollars! Let us pray...

So I think we have come full circle.

DR. DAUMAS-DUPORT: One has to consider the fact that there are different types of stereotactic biopsy procedures. Of course, a correct diagnosis cannot usually be established from only a single small biopsy specimen. In my experience, serial stereotactic biopsies measuring 10 mm in length by 1.5 mm in diameter, obtained from several locations within the tumor, which sample the different CT attenuation areas, can provide more representative sampling than a large piece of tumor obtained from a single location at open surgery.

DR. KEPES: What is the evidence from the biopsies that you have seen?

DR. DAUMAS-DUPORT: At my institution, one to three tracks are performed with three to eight biopsies along each track. Of course, depending upon the location, only one or a limited number of biopsies may be taken, but whenever possible, serial biopsies are performed. In my experience, comparison of such biopsies to tissue made available by subsequent open surgery shows the information provided by serial stereotactic biopsy sampling to be very accurate.

It is likely that in the future the indications for stereotactic biopsies will not be restricted to patients for whom surgery is contraindicated. Therefore, serial stereotactic biopsies will permit extensive preoperative histologic assessment of the tumor. Stereotactic serial biopsies establish not only a histologic diagnosis but also provide useful information regarding the geographic limits of a tumor and will often dertermine whether the tumor is composed of solid tumor tissue, infiltrated but intact parenchyma, or both. Serial biopsies can be performed very safely if based on carefully performed stereotactic imaging techniques and stereotactic angiography. Unfortunately, simple CT-guided biopsies are commonly called *stereotactic biopsies*. Given the risks, these inadequate methods allow only limited, and in my view inadequate, sampling. The concerns of pathologists are, of course, understandable when such techniques are considered. Please do not equate free-hand biopsies or single sample CT-based stereotactic biopsy with carefully performed stereotactic serial biopsies based on imaging data performed by experienced stereotacticians. They are two different things.

DR. KIRKPATRICK: Many of us are excited with the potential of stereotaxis for not only achieving a diagnosis, but also substantially lessening the morbidity for the patient.

DR. KLEIHUES: I should like to support this view very much. In our experience, stereotactic brain tumor biopsy may not be the first choice and should be restricted to those patients in whom the neurosurgeon feels unable to carry out conventional tumor resection. Only since the introduction of stereotactic biopsies have we observed a substantial decrease in the number of patients subjected to radiation and/or chemotherapy without a proper histologic diagnosis.

DR. FIELDS: At M. D. Anderson Hospital we have come to rely more and more on stereotactic biopsy. It makes it simpler and safer to get to certain locations which we could not approach otherwise. I am referring principally to the brainstem and deep forebrain structures. Today we have sophisticated imaging techniques and in a CT scan one may find a nonenhancing hypodense area, which may even occupy almost an entire frontal or, more frequently, temporal lobe. What to do?

There is no way I know of, even with magnetic resonance imaging, to make a decision with respect to the grade of the tumor. We frequently do a biopsy by CT-guided stereotaxy and may find that the specimen is all the way from showing something one cannot even say is a tumor for sure, however bad the image looks, to one that looks like a low grade astrocytoma, or to another one that is clearly an anaplastic astrocytoma. One must have a biopsy in order to make the next decision, whether to operate, to irradiate, or to just watch the patient. The morbidity of the procedures to which one may subject the patient can be modified to some extent by the use of stereotactic biopsy, which, as Dr. Kirkpatrick just said, is a method for making a diagnosis with a minimum trauma. That is a very important consideration.

DR. BIGNER: I think it is very important for this assembly, if we are going to discuss this problem, to talk about it in terms of principles of cancer treatment. Stereotactic biopsy is here to stay. Dr. Sandra Bigner has written several papers about it and makes part of her living dealing with it, and we think we can deal with it diagnostically very well. There is no question that it can provide useful information in certain settings, and I foresee it giving us an opportunity, as we gain experience with many of the more modern procedures, to even do serial samplings.

We will soon have a way to identify Mer-positive and Merminus cells in stereotactic biopsy specimens. The expression of Mer positivity and Mer negativity in tumors may shift during treatment. It would be very useful to be able to sample tumors at different points in their evolution with stereotactically obtained biopsy samples. Much of the monoclonal antibody mediated therapy is going to depend on such sampling also.

But if you are talking in terms of principles of cytoxic cancer therapy, in all situations, any kind of cancer therapy works better with the smallest possible number of tumor cells existing at the time of treatment initiation. I think it is patently ludicrous for a patient who is going to go on subsequent treatment protocols, to have a stereotactic biopsy or needle biopsy made where one is not resecting the maximum number of tumor cells present to reduce the tumor burden as much as possible prior to other treatments. This applies to any conceivable form of cytotoxic cancer treatment therapy. It is very important for us to constantly communicate that to our clinical colleagues, and keep the whole matter in perspective.

DR. SCHEITHAUER: Some of the comments that I was going to make have in part been made by a number of others already. I think that we are being asked to give the surgeon and the therapist considerably more information than tumor type and grade. Dr. Daumas-Duport brought that out in her discussion of the realationship of tumor infiltrates to the brain parenchyma. These questions include "Just how much intact tissue remains within an infiltrate:", and "Just how extensive is the tumor:" These are matters we can only resolve in a meaningful way by doing CT-guided stereotaxy.

Another fundamental question is "Is this tissue representative: We should develop, as most neuropathologists have, a basic understanding of CT and MRI scans and what they represent. If we correlate the changes with what you are seeing under the microscope, then we find ourselves in a position, even at the time of frozen section, to tell our surgical colleagues that a biopsy that shows only rare tumor cells, taken from an enhancing lesion, is not representative.

Let me digress a bit. Dr. Daumas-Duport has presented data regarding the issue of tumor grading. I come from an institution that has historically graded all neoplasms. think that we should continue to make an attempt to grade gliomas. More importantly, however, emphasis should be placed on developing grading schemes that are simple and reproducible. I am not aware of any other studies, similar to that which Dr. Daumas-Duport and I did which have attempted to compare inter-observer variability in a review of a number of astrocytomas. Grading based upon the criteria that she selected, namely, atypia, mitosis, vessel proliferation, and necrosis, when considered from the koint of view of their absence or presence, is fairly easy to reproduce.

I was surprised that Dr. Brucher was not able to find more of a correlation, and indeed found no relationship with atypia, necrosis, or endothelial proliferation. As you saw from our statistical presentation, those are highly relevant criteria for grading.

An important thing to remember about our presentation is that the study of astrocytoma grading began by excluding all favorable prognostic or *special* variants, such a pilocytic, microcystic, subependymal giant cell and pleomorphic xantho-astrocytomas. Since all of those were excluded from consideration it implies that the general pathologist, to whom this grading system is being suggested, has to be able to exclude these as well. Once they are excluded, and only diffusely infiltrative astrocytomas remain, one can see from the diagram that four distinct survival curves were generated. The one thing that I would point out is that there were only two lesions that showed absence of all criteria, i.e., only about 1 or 2% of tumors were Grade I. Three main curves remain.

Basically, I think we are all presenting the same data. It is just that with variations in grading criteria and their application, all our survival curves are a little bit different. The Grade III curve lies between the two main curves generated by applying the Kernohan grading method (i.e., a low and a high grade curve) and therefore is representative of a reproducible intermediate grade. You will remember that it is a curve that includes tumor, that show mitotic activity. If we define Grade III as a tumor having any mitoses at all, it is obvious that we are including among our Grade III lesions some of what many would call *Grade II* tumors. That principal determinant, mitotic activity, is what places the curve in that rather advantageous position. The validity of the method is supported by the large number of cases in this study. consider it a valid method. One might designate a tumor *Grade III* on the basis of a solitary mitotic figure, whereas in other instances, mitoses are numerous. We consider them both as being Grade III, but that is the cost of this kind of system.

I have already addressed the issue of stereotactic biopsy and sampling, but it is important that we try to give our clinical colleagues more information regarding the extent of these lesions. Such data is important for the purpose of planning radiotherapy. In some institutions, tumor size as determined for radiation field construction is evaluated based upon the extent of CT enhancement. We know from our studies and all of you are aware that such criteria are not adequate. The tumors are far more extensive. The process of stereotactic tumor mapping is tedious but has another useful spin-off. In operative decision-making, the degree to which intact parenchyma remains in a tumor is a very important issue. We should try to address that, even in our day-to-day practice. We have not yet published our data, but we have done a study on patients whose low grade tumors were debulked, looking at quality of survival versus the extent of tissue involvement. Significant and sometimes massive postoperative neurologic deficits were noted in patients whose tumors were sparse in cellularity, whereas the more the tumors approached a solid configuration, the better the prognosis for the patient.

These are all issues that we should address, in addition to the process of simply assigning tumor type and grade.

DR. KIRKPATRICK: With regard to radiation therapy, it seems to me that if our grading system did permit avoiding radiation therapy in some of these not benign, but more slowly growing tumors, we might be performing a service. Here in Houston we have certainly seen increasing numbers of radiation-induced neoplasms in long-term survivors. So this modality of therapy is not benign, and if by using a grading system could postpone radiation therapy until later in the life history of the neoplasm, that would be a worthwhile achievement.

DR. ARMSTRONG: We have been hearing about grading and naming of tumors by neuropathologists. A group at The University of Texas School of Public Health here in Houston has been studying the effects of environmental exposures on the incidence of brain tumors in the Texas and Louisiana Gulf Coast area. In conjunction with this study I have had the opportunity to review the diagnoses of brain tumors by community pathologists which were used as a basis for determining the incident values.

One other neuropathologist and I reviewed all of the material which consisted of glial tumors in males over the age of 21 which occurred in the target area during the study period. We identified discrepancies of diagnoses in the types of glial tumors and grade of malignancy. We encountered 40 diagnostic terms used by the community pathologists and we attempted to find the source of these terms: 18 were in the WHO classification, 18 in Zülch's book, 15 in Russell and Rubinstein's book, 14 in Burger and Vogel's book, and 4 from Kernohan's classification. (I chose these textbooks only because they have been translated into "Texan"!)

There were 15 diagnoses which were novel or unique, and in order for us to see whether that was in fact a discrepancy between the neuropathologist's diagnosis and the community pathologists' diagnosis, we devised a working classification, in which we translated their diagnosis into equivalent diagnostic terms. Using one classification of equivalent diagnostic terms, we could identify discrepancies in the diagnoses. The community pathologist made the diagnosis of astrocytoma in 219 cases, among which the reviewing neuropathologist reclassified 10 as oligodendrogliomas, 3 as other glial tumors, and 8 as other diagnoses (non-glial or non-tumor).

There was greater discrepancy between the community pathologists and the neuropathologists with regard to grading. We changed the grade from low to high grade in 37 cases. There was almost complete agreement on the diagnosis of high grade lesions in 184. It seems from our observations in this study that criteria for grading will require more consideration in any revision of the WHO classification.

DR. ZÜLCH: I am more than a little confused by this discussion on grading. What is grading? Grading is our attempt to provide information for the surgeon or physician about the probable fate of the patient.

I want to defend my friend Brucher, because the *Blue Book* tried to provide a rule of thumb prognosis by stating grades I, II, III, IV. Now, how did we arrive at those four? I can assure you that it was only after a very long discussion. Dr. Brucher showed one table from a paper by Wechsler and myself, where we tried to arrive at a definition based on survival. For example, as a result of the general feeling everybody expressed, anaplastic astrocytoma was placed in grade II. Nobody ever said what this meant.

I thought the Kernohan-Sayre classification was dead but at this meeting we have heard about it at least ten different times. Years ago when Sayre was in Cologne, we forced him to tell us what was meant by grades I to IV. He said, "Well, you find the undifferentiated cells and you count them. If they number 25% of the total population, it is I; below 50%, that is II; below 75%, it is III; 100%, it is IV. I can show you where this is written. Furthermore, he was very insistent in emphasizing that.

Today, I learned that everyone tries to provide some additional characteristics, like hypercellularity. We have all seen an oligodendroglioma which is as hypercellular as a classical medulloblastoma. We have seen today a hypercellular ependymoma. Yet ependymomas may have no mitoses, oligodendrogliomas may have a few, and medulloblastomas may have many. So we see that hypercellularity does not suffice.

The next characteristic is polymorphism. I know everybody has said here that the glioblastoma has polymorphic cells. I disagree. The fusiform cell type of glioblastoma is not polymorphpic. It is very isomorphic. Of the fusiform cells, some are longer and some are a bit shorter, but this is not polymorphism. We have always tried in our laboratory and in our group to speak of cytologic polymorphism and tissue polymorphism. We have seen cytologic polymorphism demonstrated here today. The worst of these was my favorite, the monstrocellular sarcoma.

We undoubtedly need a lot of characteristics to properly define grades I to IV. But I think that is a futile exercise because what we get these days from the surgeon is little pieces obtained under the operating microscope. My collection of tumors in Cologne comes from the time when, under the influence of Dandy or Olivacrona, lobectomies were performed. I had the advantage of examining large paraffin sections of these tumors in which you would see the common variations. I have always tried to emphasize these variations. With stereotactic biopsy, if one does not get to the right place, he may be in great difficulty.

I really think we should not go too far in this attempt to define grades by such characteristics, but rather go the other way. Lucien Rubinstein's presentation was a very fine example. If he had undertaken classification and grading only according to the morphology, there would have been no correlation with survival time. We always took the other path. We tried to find a series of tumors which were similar in morphology, whatever that is, and then tried to make a correlation for the neurosurgeon by asking him his impression regarding localization, size of the mass, and his estimation of the survival time? Then we had rule of thumb definitions of grading as was expressed by Brucher. We thought that this was a quite reasonable approach.

DR. KLEIHUES: In my view, the cases presented by Dr. Rubinstein are very convincing. In the context of our discussions on the concept of primitive neuroectodermal tumors, his study is also a very timely one. I must admit that the ependymoblastoma is one of those tumor entities in which, until now, I had little faith. Specifically, I would like to know whether the presence of true ependymal rosettes is a consistent feature throughout these tumors.

DR. RUBINSTEIN: It was not a consistent feature. Sometimes there were very few ependymal rosettes, sometimes there were typical perivascular pseudorosettes, and sometimes the tumors had a cellular epithelial pattern, but everybody who looked at these tumors agreed on the diagnosis, particularly on the basis of the cell nuclei which are so very characteristic in ependymomas. So, while there was no difficulty with the diagnosis, there was no particular correlation with the number or frequency of rosettes.

However, the features of anaplasia were invariably present: i.e., the loss of structural pattern, the necroses, the vascular endothelial proliferation, and especially the mitotic activity. All except one of these cases had been referred to us because they were recognized by the referring pathologists as malignant, and their prognostic assessment was both important and difficult.

DR. KLEIHUES: It is a rare entity.

DR. RUBINSTEIN: It is difficult to say. A recent review by LEIBEL and SHELINE (1987) states, on the basis of various series, that about 20% of ependymomas are malignant. However, some of these series undoubtedly include medulloblastomas. One suspects this because some series, like those reported by SALAZAR et al. (1975) state that they have identified more malignant ependymomas than benign ones, which is not the experience of others. So there must be some misdiagnoses here. In addition, the 20% figure collated from all of the series, was arrived at without drawing a distinction between ependymoblastomas and malignant (or anaplastic) ependymomas. The distinction was, for instance, specifically ignored in the series of SHAW et al. (1987).

DR. KLEIHUES: Is there a clear single criterion by which one can to distinguish the malignant ependymoma from its embryonal counterpart, the ependymoblastoma?

DR. RUBINSTEIN: Yes, the distinction is usually clear-cut.

DR. KLEIHUES: The second paper I would like to address is that of Peter Burger on the grading of gliomas. Our laboratory was chosen as the pathology reference center for a German/Austrian glioma therapy trial. Included in this study were patients with anaplastic astrocytomas (WHO grade III) or with a glioblastoma multiforme (GBM). The correlation between morphologic grading and patient survival revealed two interesting aspects. First, a significant difference in survival time between patients with anaplastic astrocytoma and glioblastoma multiforme was only observed on the basis of grading as established and carried out by the pathology reference center. When survival curves were plotted according to the grading by the referring pathologists, a significant difference in survival was no longer detectable, indicating that the criteria used are highly divergent among practicing neuropathologists.

In addition to anaplastic astrocytoma (WHO grade III) and glioblastoma multiforme we established a grade III-IV tumor which we defined as a neoplasm which in large tumor areas still shows the typical appearance of an anaplastic astrocytoma whereas in other tumor portions the morphologic criteria of the glioblastoma are already fulfilled. Formerly, such tumors were termed secondary glioblastomas. An analysis of survival showed that patients with these tumors did as poorly as glioblastoma patients, at least during the first 18 postoperative months. Long-term follow-up, however, revealed that some of these patients (less than 10%) show increased survival. On the whole, this study confirms that when the criteria for the diagnosis of glioblastoma are focally present anywhere in an anaplastic astrocytoma, tumor biology and clinical outcome are very likely to correspond to those of the glioblastoma.

DR. KIRKPATRICK: Dr. Brucher, is what Dr. Kleihues has said about the focality of glioblastoma not affecting the diagnosis inconsistent with what you have said earlier?

DR. BRUCHER: Yes, I want to make the point clear. The problem is, when you are looking at diffuse astrocytoma and not a glioblastoma, you have to recognize the presence of anaplastic foci. If your criteria are pleomorphism, necrosis, and endothelial proliferation, you will not recognize anaplastic foci in an astrocytoma.

My method is to look with the low power microscope in order to recognize hypercellularity, which is a consequence of proliferation. In those foci of hypercellularity there will develop mitotic figures, and less differentiated cells. If you recognize the three criteria together, you may be almost sure it is an anaplastic focus, and then, I will give it the qualification of anaplastic astrocytoma. My problem was not an inability to make a distinction between anaplastic astrocytoma and glioblastoma.

If the sense of this discussion is "what is the difference between glioblastoma and anaplastic astrocytoma", then both necrosis and endothelial proliferation will be important chacharacteristics. I agree with you but the list of the three important histologic criteria concerns only the recognition of anaplastic foci in the diffuse astrocytoma.

DR JÄNISCH: I want to remind you that Dr. Sobin made some remarks yesterday concerning grading, and he reminded us that there are two types of grading. One is histologic grading, which means actually a grading of anaplasia. As Dr. Zülch mentioned today, anaplasia concerns the cell or the tissue. Another type of grading is biologic grading. We heard from Dr. Burger and we know from the literature that in brain tumors, of course, other factors such as histologic grading affect the individual prognosis. Most of these factors are outside the realm of possibility for the pathologist. If you are in close contact with the clinician, or if you have the opportunity to see the patient yourself to study in detail the history of his disease, then the pathologist also can arrive at a biologic conclusion.

In practice, at least in our country, where we are often a considerable distance from the neurosurgical clinic, I can not tell in any given case what the Karnofsky performance of the patients was, and so on. The best information that I can obtain is the age and the sex of the patient and maybe some short notes on the anamnesis. Therefore, as a pathologist, I have no right to make such a biologic grading for a single patient, for a specific patient. What I must do is make a morphologic grading and put it in the hands of the clinician. Then, he has to consider it in context with all the other factors influencing the prognosis of the patient.

The curves which have been discussed here are all statistical values. They are drawn from between 100 and 200 patients so this does not mean that a patient who is in grade I will live five years or longer; he may die sooner under circumstances which depend on the biologic process, the behavior of the tumor and its location. In another case, a patient with a grade III tumor might live five years or longer. Therefore, I think we have to stick to the anaplasia grading, the morphologic grading, and not go into details regarding how long a particular patient with such a tumor will live since this also depends on the treatment.

For these reasons, I would support the view expressed by Dr. Burger, that this grading system which we already have in the Blue Book could be a sound basis for further discussion. We must make the definition more precise so that we understand what is meant by grade I from the histologic or morphologic point of view. Dr. Rubinstein's presentation reminded us very well that we cannot have the same grading criteria for all tumors. We have to define this more precisely for each tumor type.

DR. JELLINGER: The first multicenter study by a German group was in the mid-1970s (Kuhlendahl et al., 1973). It included only supratentorial gliomas. The Kernohan grading I to IV was used. Histopathology was reviewed by several neuropathologists. There were two nonrandomized groups of patients, who after surgery were treated with or without conventional radiotherapy and without chemotherapy. Between these groups, we had about 50% difference in the mean survival time. In an independent study comparing group 3 with group 4, there was very little difference in survival time between grades, irrespective of treatment (Grunert et al., 1973).

These data were in contradistinction to the later American studies which very often combined patients with grades III and IV.

One of the most important observations of more modern studies is the difference in biologic behavior of grades III and IV. However, we have to keep in mind that even in the grade IV group, i.e. the glioblastoma group, there are about 5% with long survival. We believe that one of the advantages of more aggressive treatment is that the number of long-term survivors is increasing.

DR. SCHEITHAUER: There are a lot of neurologists who have patients with CT-hypodense lesions whom they have followed for years. Their position is: "We don't really want to do anything. The patient is neurologically intact." These lesions are in most cases infiltrative gliomas and the hypodense area will enlarge with time. A T-2 weighted MRI image looks even worse than the same lesion when seen on CT. Radiotherapists make a very valid point--if they are going to be successful in eradicating a tumor, their best shot at it is when it is in its early phase of growth.

I can understand why a neurologist might not want to have that tumor debulked. There would be a devastating deficit in some cases, depending on tumor location. But I suscribe to the viewpoint of the radiotherapists. With a diagnosis of low grade infiltrating astrocytoma of the diffuse type, not a pilocytic tumor or some other favorable lesion, the patient should be irradiated. Our data supports that view.

DR. RUBINSTEIN: On what basis should they be irradiated? You are assuming here that there are very few neoplastic cells. Therefore, the growth fraction of the tumor, if indeed it is a tumor, must be very low. What are the chances then that the lesion would even respond to radiation? If the patient has no symptoms, why irradiate, and what is the rationale?

DR. JELLINGER: Who can exclude some anaplasia induced by irradiation? This is an unresolved problem.

DR. YUNG: I think we really do not know what we are doing to the patients. A patient may have a hypodense area seen on the CT scan which probably represents a very low grade astrocytoma. We do not know the natural history of such a lesion.

Dr. Kleihues mentioned a series of treated patients who progressed from low-grade astrocytoma to anaplastic astrocytoma to glioblastomas. We do not know when we see a very low grade astrocytoma, whether it is going remain unchanged or undergo anaplasia, if we leave it alone.

DR. KEPES: Needle biopsy is a very sensitive issue. When the choice is no biopsy versus needle biopsy, the latter is clearly preferable. My point is that needle biopsy should be a starting point. Suspicious or positive lesions can then be followed up. Unfortunately, it has become the final word in some places.

DR. FIELDS: Lucien Rubinstein really got to the heart of the matter. Another one of my patients was a 28-year old female Ph.D. computer engineer for IBM, who had a left frontotemporal lesion and no neurologic deficit. After surgery and radiation, she has had sufficient intellectual impairment to influence her decision-making in her job. That is the kind of problem that I have to face as a clinician. That is what worries me about treating these patients too vigorously.

DR. RUBINSTEIN: I would have recommended leaving her alone.

DR. KLEIHUES: We have recently performed a postmortem study on a patient who suffered from chronic, therapy-resistent epilepsy for 28 years. During the last 20 years of her life she was in a hospital for chronic epileptics. We found a very well differentiated slow-growing oligodendroglioma of the temporal lobe and assumed that this benign lesion was responsible for epileptic seizures extending over a period of almost three decades. In my view, early surgical intervention might have cured her. Radiation therapy, on the other hand, would probably not have been beneficial to the patient.

DR. KIRKPATRICK: That may not be an exception.

DR. FIELDS: I fail to see how you can put oligodendroglioma in the same context, at least clinically, with an astrocytic tumor.

DR. KLEIHUES: I get the impression that many radiotherapists are changing their attitudes. Only a few years ago they would irradiate every glioma, including pilocytic astrocytomas of the cerebellum and the optic nerve. Now they have become more hesitant in applying radiotherapy to highly differentiated gliomas. The basic problem is that until now, a prospective randomized study on the effects of radiation on low-grade gliomas has not been conducted. Therefore, two European studies are presently being initiated by the European Organization for Research and Treatment of Cancer (EORTC) and another by the German/Austrian Glioma Group.

DR. KIRKPATRICK: Just to re-emphasize, that yes, we suspect also that radiation therapy is superfluous in the low-grade astrocytomas.

With regard to epilepsy, I would like to just cite Dawna Armstrong's very nice study with Robert Grossman and others at Baylor College of Medicine here in Houston, on temporal lobectomies for epilepsy, showing a really substantial number of benign neoplasms. A lot of them are gangliogliomas and oligodendrogliomas.

DR. BURGER: The Brain Tumor Study Group has just started a randomized study and there are two arms. All patients get surgery, one gets radiation and the other does not. There is considerable unease among neuro-oncologists that these patients are going to be irradiated, and then survive long enough to develop an encephalopathy. I think the incidence of radionecrosis is probably fairly low, but there are a number of neuro-oncologists who think that anyone that gets 5,000 rad, and survives more than two years, is at risk. Even 4500 to a local area may be enough. So there is considerable concern that this is going to produce problems.

DR. FIELDS: I, for one, certainly share that concern. Radiotherapy is not, by any criteria, a benign modality of treatment.

DR. DAVIS: By means of summary, I do not think I would be terribly unkind if I characterized the presentations today as varying from the sublime to the semantic. We began this morning with Darell Bigner presenting a very interesting, though preliminary, presentation on studies of karyotyping of malignant gliomas. He reiterated a number of terms that are characteristic of this field, *deletions* and *gains*, *double-minutes*, *gene amplification* and, of course, *oncogenes*. He told us that some of the procedures were going to become automated. He talked to us about perfect cancer cells that turn themselves on, and told us that some of these products are related to Drosophila genes. How that relates to brain tumors, we are not sure yet. However, we are sure that the abnormalities are really very difficult to find. He talked about "looking for a needle in a haystack"; I would say "looking for a locus in a gene stack."

Paul Kleihues then presented us with some work using some of the new marker products, and indicated that there was a higher degree of reliability of diagnosis by using marker products, but that they were not helpful in determining the degree of anaplasia. He stated that the new markers for BrdU and Ki-67, in particular, gave insight relative to the proliferative capacity of tumor cells (HOSHINO et al., 1988). He also showed us the correlation of morphologic features and markers, NSE, synaptophysin and neurofilament protein, in medulloblastoma.

In the discussion of these two papers, it was pointed out by Dr. Rubinstein that the markers offered possible insight into familial gliomas, for which there is a registry in Leyden. I was unaware of this registry, and I thank him for that information.

Then, Professor Zülch reminded us that there was history of brain tumors before Bailey and Cushing. He emphasized the importance of tissue relationships and knowledge of variants of brain tumors. He gently chided us for not accepting the pineal parenchymal origin of some germinoma-like lesions, or his favorite tumor, the monstrocellular sarcoma.

There then followed a series of presentations by Dr. Daumas-Duport, Dr. Burger, Dr. Brucher, Dr. Wechsler, and myself, circling around the problem of the nomenclature of the gliomas, particularly relative to anaplasia. I am not sure that we got all the way around, or at least I was not sure about that until Dr. Scheithauer, in his summation, talked about where the different curves lie.

There was clear internal consistency within each of the varying gradation schemes, but there still are variations in definition that do not allow a good comparison between different studies. I truly believe that this discussion brought out the need for a universal classification if we are going to get meaningful comparison between the different groups.

Dr. Rubinstein gave us an exposition in which he separated quite clearly the ependymoblastoma from the other so-called malignant ependymomas. He indicated that in the latter group, at least, there was not a uniformly poor prognosis.

During the discussion of the last series of papers, there was re-emphasis of the problems of definition, of microscopic appearances with all sorts of stains, with all sorts of markers, and these are variously useful in helping us determine the kind of tumor and the degree of anaplasia.

I was interested that there had been a translation of a number of textbooks into Texan. That has not happened yet in my home state. There seems to be a great deal of difficulty on the part of many individuals in reading any of the textbooks because I do not think that they read them carefully and translate them into reliable repeatable diagnoses.

DR. KIRKPATRICK: Thank you, Dr. Davis. As the author of a distinguished, unread textbook, I am sure you know what you are talking about.

If we are supposed to reach a consensus, I think that goal is just a little bit presumptuous. Unless there is a substantial sentiment to endeavor to reach a consensus at this point, we will have to continue our thinking and talking about the necessity for a realistic grading system, which recognizes the continuum of malignant change in these neoplasms. We will also have to consider a few of the low frequency astrocytomas which have not been mentioned, such as the astroblastoma and the pleomorphic astrocytoma. Those need to be included somewhere in the classification.

PART THREE

MENINGIOMAS

Moderator:

JANET M. BRUNER

CHAPTER XVI

History and Diagnosis of Meningiomas

JOHN J. KEPES, *Department of Pathology, University of Kansas Medical Center, Kansas City, Kansas*

It gave me special pleasure to meet here in person for the first time Professor Klaus J. Zülch, the author of the first edition of the WHO *Histological Typing of Tumours of the Central Nervous System.*

None of us neuropathologists gathered here developed our knowledge and understanding of the subject in a quick and easy way. We all had to start the learning process somewhere. For me, that time was about 35 years ago when I joined the newly founded National Institute of Neurosurgery in Budapest Hungary, as a pathologist with background only in general pathology but not in neuro-oncology. The director of the Institute, Professor Zoltan, fully understood my problem and suggested that I start studying the elements of CNS tumor pathology from the original German version of Professor Zülch's book on brain tumors. He gave me six weeks to learn the contents of the book before I started looking at operative specimens at the new Institute. Although naturally I have learned a great deal from other teachers and printed sources since, Professor Zülch's book was the first compass used during my journey into neuro-oncology. For this reason I would like to dedicate this presentation to him, or as they say in German: *"Diese Vorlesung ist Herrn Professor Klaus-Joachim Zülch gewidmet!"*.

This meeting is about nomenclature and histologic variants of tumors of the central nervous system. My subject today is meningiomas. The foregoing discussions of glial and other neuroectodermal tumors have amply shown that the crystallization of today's views on these neoplasms took a long time, but I submit that with meningiomas we have come an even longer and more arduous way. At least with gliomas, the fact that they represent the neoplastic form of *glial* cells was recognized even in the 19th, and certainly by the early 20th century, with terms such as *sarcomas* and other inappropriate names for glial tumors having been discarded a long time ago. Also, there was never much doubt about the fact that the great majority of gliomas started their growth within the brain or spinal cord.

With meningiomas even the gross anatomical origin of those neoplasms was mistakenly believed to be the *dura mater*, *fungus durae matris* having been one of the original names applied to these neoplasms (LOUIS 1774) and it was a long uphill struggle before the observations first by BRIGHT (1831), and later by CLELAND (1864) and SCHMIDT (1902), about the similarity between the cells of meningiomas and those of arachnoidal cell clusters convinced most observers that these tumors are indeed derived from arachnoidal cells. Since the basic character of the latter was also shrouded in mystery, it should not be surprising that rather than recognizing in name also the unique characteristics these cells possess, for the tumors of the arachnoid terms were borrowed from general pathology that were by necessity inappropriate. Thus, in addition to the gross descriptive term of a *fungus* meaning a fungating mass and *psammoma* (VIRCHOW 1864) (a term that really deals only with one isolated feature of these tumors and is not even present in many of them), others thought to recognize features of carcinoma, sarcoma, fibroblastoma and endothelioma in these neoplasms and until well into the 1930's such terms were often used to describe meningiomas (PENFIELD 1932). It is a fact however, that the arachnoidal cap cells cannot properly be described as epithelium, fibroblasts or endothelial cells. If anything, DEL RIO HORTEGA's suggested term *exothelioma* (1930) seems more appropriate because meningothelial cells are not *lining* but *covering* from the outside that layer of the leptomeninges we call the arachnoid membrane. The use of terms like *carcinoma, sarcoma, fibroblastoma*, etc, for meningiomas amounted to trying to drive the proverbial square peg into various and sundry round holes. It was time to recognize the uniqueness of this neoplasm just as in another field of oncology the name *melanoma* has replaced the earlier terms *melanocarcinoma* and *melanosarcoma*, and rightly so, since melanocytes are neither true epithelial cells nor fibroblasts.

As we all know, in his 1921 Cavendish lecture, Harvey CUSHING (1922) cut through the Gordian knot by proposing the name *meningioma*, the term that has served us well for the last 67 years, and one that is not likely to be replaced in the near future. Nevertheless, as Drs. RUSSELL and RUBINSTEIN (1988) pointed out in their book some detractors even at that time felt that Dr. Cushing took a "forward step in a retrograde direction."

Well, what if anything, could be wrong with the term *meningioma*? In my opinion its principal weakness is that it substitutes a gross anatomical structure, the meninges, for a more specific name of tissue or cell groups as being the origin for these neoplasms. To use an analogy from general pathology, it is as if one called mesotheliomas of the pleura or of the pericardium simply *pleuromas* or *pericardiomas* respectively, or referred to a synovial tumor originating from a bursa or a *bursoma*. Although admittedly a little cumbersome and unlikely to attain much popularity for tumors of arachnoidal cell origin. Del Rio Hortega's suggested term *meningoexothelioma* was a good one in terms of exactitude, although *meningothelioma* would also be quite acceptable.

As you all know very well, the histologic patterns of meningiomas include variants that suggest a *talent* for mesenchymal differentiation and others that point to *epithelial* behavior. Two explanations alone or in combination, might be invoked to help understand such erratic behavior. One obviously would be that the meninges develop from both mesenchymal and neuroectodermal elements.

I had the pleasure and honor to have as a speaker at the Presidential Symposium on Meningiomas at the 1985 meeting of the American Association of Neuropathologists in Boston, Professor O'Rahilly, the curator of the Carnegie Embryological Collection. He is a person eminently qualified to report on early embryologic development of any or all of our organ systems. On that occasion he pointed out that the human meninges are derived from several types of tissue elements including neuroectodermal as well as mesenchymal ones but these are so smoothly blended together that in the mature meninges, no histologic distinction is possible between areas like the frontal lobe meninges that are predominantly derived from the mesoderm and others like the lower spinal meninges that are mostly of neural crest or tube origin (O'Rahilly and Müller 1986). But does embryology provide the only explanation for the diverse behavior and potential of neoplastic arachnoidal cells? I for one certainly agree that not only will structure play a role in function (Rubinstein 1986), but the reverse is also true, namely, that the assigned position and *job* of certain cells, regardless of their embryologic beginnings may invest them with structural features that in turn enable them to perform adequately in their respective posts. Here, I would like to use once again the example of the mesothelial and the synovial cell, neither of which, as far as I know, is regarded as anything but mesenchymal in its origin. Yet, since they are positioned in a quasi-epithelial arrangement (one to cover serous membranes, the other to line joints, tendon sheaths and bursae), they do display, as is well known, both epithelial (papillary and adenomatoid) as well as fibroblastic features in their tumors. This being the case, is it difficult to decide how much of the dual (epithelial and mesenchymal) *talents* of meningioma cells are derived from the mixed ancestry and how much of them are role related?

Since this is the age of immunohistochemistry, it is perhaps proper to stress at this point that several immunoperoxidase studies, best exemplified and summarized by Meis et al. (1986) in their study of 50 meningiomas, found both *epithelial* and *mesenchymal* type antigens to be regularly present in these tumors.

Many features of the mesenchymal potential of meningothelial cells have been known for a long time, although, until the advent of electron microscopy, there were some who felt that fibroblastic meningiomas originated from the *stroma* of the arachnoid, whereas the meningothelial forms came from the cap cells, a view no longer held.

You have all seen examples of the well-known *mesenchymal* variants of meningiomas; fibroblastic, lipoblastic, myxoid, chondroblastic, osteoblastic (as distinct from hyperostosis involving the overlying skull) and xanthomatous forms. The term *angioblastic* is unfortunately often used very loosely to stand for what CUSHING and EISENHART (1938) clearly meant to represent only one of the three subgroups of their angioblastic meningiomas, namely the hemangiopericytic type. The other two are leptomeningeal examples of Lindau's tumor (i.e. hemangioblastoma), and what the WHO classification refers to as *angiomatous meningioma* (ZÜLCH 1979). The latter is understood by most as being an otherwise ordinary meningothelial meningioma that happens to have a great many blood vessels. Allow me to comment first on hemangioblastomas of the meninges. They do exist either together with other manifestations of von Hippel-Lindau disease, or independently from them. They are, in my opinion, not true meningiomas at all. Like their cerebellar or brain stem and spinal cord counterparts, they do possess a dense capillary network with stromal cells (sometimes heavily lipidized) between them, but these stromal cells do not have characteristics of meningothelial cells. For example, they form clusters between capillaries but do not wrap themselves around the vessels in a concentric form, just as they refrain from doing so in Lindau tumors seen in other locations.

As to *angiomatous meningiomas* as stated above, some of them are simply highly vascularized versions of meningothelial meningiomas, but in some of them, and perhaps these tumors ought to be designated by a different name, the relationship between meningothelial cells and capillaries is close and intimate to an extraordinary degree including the presence of one or more capillary channels between meningothelial tumor cells in many instances (KEPES 1982). In these tumors lipid droplets, probably plasma lipids derived from leaky neoplastic vessels, often fill tumor cells. Moreover, nuclear pleomorphism and hyperchromasia, for reasons unknown to me, is often encountered in such tumor cells without, I believe, indicating anaplasia or malignancy since other histologic features, such as mitoses or necrosis, are not encountered. Clinically as a rule these tumors behave in a benign fashion. One intriguing question remains, namely, is there such a thing as a truly *angioblastic* meningioma in the sense that meningothelial cells themselves become transformed into vascular endothelium? This is clearly not the case in *routine* angiomatous meningiomas or in the meningeal forms of Lindau tumor, or even in hemangiopericytomatous meningiomas (where it is the *pericyte* and not the usually benign-appearing vascular endothelium that is regarded as the neoplastic element). But there remain a few meningeal tumors where the possibility of this transformation is perhaps still open. ERMEL (1974) for one, believed that by electron microscopy he could trace transitional forms between meningothelium and vascular endothelium. In our material such transition could not be substantiated so far.

The problem of the nosologic standing of *hemangiopericytomatous meningiomas* or if you wish, *meningeal hemangiopericytomas* is one that could fill an independent lecture and this audience is well aware of the pros and cons of the two positions. I personally belong to the minority who prefer to think of these neoplasms as still another variant of meningiomas. In that respect, I am in good company with HORTEN et al. (1977). To the well-known arguments favoring meningothelial origin for these tumors including observations of transitional forms and immunohistological findings, I would like to add three additional points. 1) Why would it be so difficult for neoplastic meningothelial cells that are able of turning into collagen-forming fibroblasts, chondroblasts, osteoblasts, lipoblasts, etc, to undertake pericytic differentiation? 2) The by now well-known *papillary meningioma* (LUDWIN et al. 1975) may be regarded as a bridge between meningothelial and hemangiopericytic meningiomas since papillary formations may develop in both types of tumors. 3) Some hemangiopericytic meningiomas undergo changes not recorded in soft tissue hemangiopericytomas, such as gradual pseudoglandular transformation which we have observed in a parietal convexity meningioma of an eleven year old girl (KEPES et al. 1983). Finally, for the pleasure of biostatisticians, I would submit that if these tumors were simply derived from whatever pericytes are normally present around meningeal capillaries, one ought to find it strange that the leptomeninges with a total weight of about 30 grams should develop as many as 9% of hemangiopericytomas, with the remaining 91% involving other parts of the body (based on data of the Kansas Bureau of Health Statistics). Since an average adult weighs about 70 kg (70,000 grams), the incidence of hemangiopericytomas in the meninges, by the most conservative calculations, is at least *100 times* higher than the relative weight of the meninges would seem to warrant.

While the above listed forms of mesenchymal differentiation in meningiomas have been known for a long time, some recent data has accumulated which further accentuates the mesenchymal potential of neoplastic meningothelial cells. As RUSSELL and RUBINSTEIN (1988) point out in their book that Marshall had stated in 1956 that cells of the arachnoid belong to the reticuloendothelial system. Phagocytic properties in meningothelial cells, for example, can be observed in some of the more common meningiomas. Here, apparently because of episodes of previous intratumoral hemorrhage, deposits of hemosiderin may accumulate not only within the fibrous septa of the tumor but as Perls' stain indicates in some such cases, the *tumor cells themselves* have also incorporated much hemosiderin. One might mention that in contrast neurons and astrocytes in areas of old hemorrhage usually become only incrustated with iron on their outer cell surfaces.

Dr. Chen and I had the opportunity in Taiwan to observe two patients with meningiomas who at some time in the past received intravenous polyvinylpyrrolidone (PVP) as a plasma expander used during abdominal surgeries and on other occasions when the drop of blood pressure was a problem. We do not use this substance in the United States for this purpose, but in countries where its use was at one time widespread, it has been observed that PVP becomes stored exclusively in cells of mesenchymal origin such as fibroblasts, histiocytes, Kupffer cells, renal glomeruli, bone and cartilage, etc). This condition was nicely described in Taipei by Kuo and Hsueh (1984), as *mucicarminophilic histiocytosis* since in those cells that stored this material (PVP), the latter provokes the accumulation of a mucoid substance which in turn causes distention of the involved cells. The meningiomas of our two patients who had in the past received intravenous PVP, showed the tumor cells to contain large amounts of mucoid material as seen on H&E stain but better appreciated with mucicarmine and colloidal iron stain. They also contained particulate polyvinylpyrrolidone as brown amorphous granular deposits (Chen et al. 1985). This phenomenon certainly firmly establishes meningothelial tumor cells as having mesenchymal capabilities in storage diseases involving the reticuloendothelial system.

Another recent observation concerns the presence of rheumatoid nodules within a tentorial meningioma in a patient with longstanding rheumatoid arthritis. In this case the meningothelial tumor cells took part in forming the rhematoid nodules themselves (Kepes et al. 1986).

When it comes to more *epithelial* behavior of meningiomas, in some areas of the neoplasm cells may be seen neatly grouped as cuboidal epithelium and of course by EM they have well-defined desmosomal attachments which are characteristic of both epithelium and meningothelial cells.

Naturally, the first morphologic feature with which everybody identifies meningiomas is the *meningothelial whorl*, with or without calcification. This tendency of covering other cells in a concentric fashion that Dr. Rubinstein likes to call *lepidic*, is recognized by second year medical students and makes it quite easy to recognize meningiomas as long as there are only a few such tumors for them to examine. With age and experience, at least in the case of meningiomas, come not fewer but more numerous problems. At any rate, meningothelial cells will form whorls in the non-neoplastic state as well as in tissue cultures and in some tumors they may comprise just about every part of the neoplasm in certain areas, whereas elsewhere the cells may become more elongated as in the so-called transitional meningiomas. It is rather interesting to observe at times how meningothelial whorls enclose within their centers other types of tissues such as striated muscle fibers as seen in the temporalis muscle invaded by an otherwise benign meningioma.

Now we also come to the question of whether to subtitle certain meningiomas as *psammomatous* because they contain psammoma bodies?. The number of such structures may range from very rare to overwhelming and occasionally focally as the exclusive component (the latter perhaps seen more often in meningiomas of the spinal canal). It may be a good idea, if you are going to retain the subgroup of *psammomatous meningioma*, to specify how many such bodies one should find per high power field before using this term?

Another *epithelial* characteristic in meningiomas which has come to be regarded, prognostically, as an ill omen is a tendency for the formation of papillary structures. Looking back on Dr. Cushing's experience, his famous piano teacher patient, Dorothy May Russell, the last patient described in great detail in his 1938 monograph with Dr. Eisenhardt, had just such papillary formations develop in the late recurrences of her meningioma (there were 17 by number). The article by LUDWIN et al. (1975) and the WHO Blue Book (1979), in my opinion, rightly single out this histologic feature as an ominous one. Needless to say, similarities between papillary meningiomas and metastatic papillary carcinomas, papillary ependymomas or even astroblastomas may cause serious diagnostic difficulties, particularly if only small pieces are available for examination. Still another quasi-epithelial feature documented increasingly these days in meningiomas, is the formation of hyaline inclusions or pseudopsammoma bodies. BRIGHT (1831) was the first to describe and illustrate these structures and then none other than Harvey CUSHING who in 1913 operated on a clivus meningioma of a certain Josephine T and exactly 25 years later included in his monograph a picture of *peculiar glassy hyaline inclusions* in this tumor. I was still working at the National Institute of Neurosurgery in Hungary, back in the early 1950's when I encountered some meningiomas with these inclusions. The inclusions were quite prominent and round, but did not develop from hyalinized whorls and certainly never seemed to calcify and so I thought that the term *pseudopsammoma body* would be a fitting one in referring to them. My original publication (KEPES 1961), depicting the light microscopic features of these structures, was followed by studies by others (BERARD et al. 1978; FONT and CROXATTO 1980; BUDKA 1980), all adding some interesting histochemical and later immunochemical observations about these structures. From the beginning it was obvious that PAS is a very good stain to visualize as many pseudopsammoma bodies as possible in a given area that had them. To this were added later mucicarmine (FONT and CROXATTO 1980), positive histochemical tests for basic proteins, and the discovery of carcinoembryonic antigen, IgG, IgA, secretory components and other substances (BERARD et al. 1978) within the hyaline inclusions. Though the cells that contained them showed epithelial membrane antigen and cytokeratin in one and the same tumor, cells without hyaline inclusions stained positively for vimentin. This we now know is the principal intermediate filament for meningothelial cells, normal or neoplastic (HALLIDAY et al. 1985). Because of the difference between ordinary meningioma cells and the ones that contain pseudopsammoma bodies, KOCK and TEGELBJAERG (1981) went so far as to suggest that these cells represent an *alien population* within meningiomas.

Ten or more years ago (KEPES 1975) it was shown by examination with the electron microscope of a meningioma containing such inclusions that the bodies were housed in intracellular lumina lined by microvilli. Such intracellular lumina were found useful for diagnosis of breast carcinoma (MICHAUD and GAGNE 1983) and later were seen in many other carcinomas as well as in cells of mesotheliomas (WANG 1974). KUBOTA et al. (1982) showed us the earliest forms of these bodies in meningiomas, consisting of tiny intracellular slits into which microvilli protruded but without any secretions in the early phase.

It has been suggested (ALGUACIL-GARCIA et al. 1986) that tumors with pseudopsammoma bodies should be recognized as a distinct subgroup of meningiomas with the name *secretory meningiomas*. I had the opportunity to examine recently with the electron microscope a number of additional meningiomas of this type, focusing on the question: "exactly what type of *secretion* are we talking about in these cases?" These recent studies allowed us to demonstrate for the first time that more than one cell may participate in surrounding a single pseudopsammoma body. This can easily be recognized from the desmosomal attachments between two neighboring cells, both projecting microvilli towards the inclusion body. Further, to our surprise we found that some of these cells may have similar microvilli on their *outer* surfaces as well, not facing the inclusion body but instead facing other not similarly endowed *ordinary* meningioma cells.

As to the nature of the secretory process, if you can call it that, some of it results in a fine granular material apparently spewing forth from the tips of microvilli which themselves are filled with the same kind of granular substance; but equally important are the stacked-up laminated membrane fragments first observed by BERARD et al. (1978) that on closer look proved to be products of desquamation from the membranes forming the walls of the microvilli themselves. In some instances, round membranous profiles and stacked up membranes comprise the entire hyaline inclusion. If what we are witnessing is indeed a secretory phenomenon, of what type would it be? Perhaps the term *merokrine* may be justified since in addition to emptying secretory material into the lumen, parts of the villi themselves become mixed up with the mass of the hyaline inclusion. This continuity between microvilli and the inclusion bodies themselves can be readily observed on scanning electron microscopic study of pseudopsammoma bodies.

An additional twist to the story of secretory meningiomas developed when CHALLA et al. (1980) and independently MIRRA and MILES (1980) discovered meningothelial meningiomas that in addition to having many pseudopsammoma bodies also contained an unusual perivascular proliferation of pericytes. Until Dr. Challa's observation no apparent clinical significance could be attached to the finding of pseudopsammoma bodies in a meningioma except perhaps that the overwhelming majority of such tumors is apparently histologically benign (although recently we have seen a case where at least the marrow spaces of the overlying skull were invaded by meningioma cells that contained hyaline inclusions). CHALLA et al. (1980) also found that these otherwise benign meningiomas often have unusually marked perifocal edema in the neighboring brain. Such was the case in Mirra's patient too, although she did not at the time attach too much importance to the massive edema surrounding her patient's brain tumor. One of our very recent cases of meningioma with pseudopsammoma bodies shows very striking edema around a relatively small sphenoid ridge meningioma. Whether this is related to the pseudopsammoma bodies themselves or to the pericytic proliferation is not known at this time, but I would like to mention that there was a striking amount of IgG found in the pericytic zone in several of our more recent cases and in one of them massive secretory activity was highlighted by greatly distended ribosome-studded rough ER systems in the proliferating pericytes, similar to that seen in active fibroblasts, but without any significant increase of collagen in the area. So these cells may well have been engaged in the production of some other, so far unknown, type of product.

In summary, I do believe that Dr. ALGUACIL-GARCIA's (1986) recommendation to use the term *secretory meningioma* as a name for this subgroup is justified on both morphologic and clinical grounds.

Meningiomas may be quite difficult to differentiate from other types of neoplasms in certain instances, of which I would like to present a few examples. Most of these are probably quite well known to you.

As was pointed out in BURGER and VOGEL's text on *Surgical Pathology of the Central Nervous System and Its Coverings* (1976) in some meningiomas tumor cells may undergo hydropic degeneration and become very similar in appearance to oligodendroglioma cells.

This has a modicum of importance since many *real* oligodendrogliomas are also superficially located and may even invade the dura mater.

Meningiomas may mimic astrocytomas when they are loosened by edema and their cells become multipolar and spidery. Conversely, eosinophilic hypertrophy of the cell bodies of meningothelial cells may bring about an uncanny resemblance to gemistocytic astrocytes. This presents a problem that can be difficult to resolve until and unless one finds cellular whorls formed by these tumor cells, but since astrocytomas at times may themselves form cellular whorls, this is not always sufficient to distinguish these two tumors. Fortunately such neoplasms if they are of meningothelial origin, will be negative on immunostaining for GFAP whereas *real* (astrocytic) gemistocytes are GFAP positive. Palisading of cells in meningiomas may come close to imitation of Schwannomas a phenomenon recently mentioned by SOBEL and MICHAUD (1985). Masson's *forme humid* or humid form should be mentioned here (also referred to as microcystic meningiomas (SOBEL and MICHAUD 1985). In these usually quite benign meningiomas intra and extracellular lipids and glycogen create a loose appearance, that in my experience is very often the cause of misdiagnosis and of requests for mailed-in consultation. Careful examination usually allows the detection of loosely structured meningothelial whorls in these tumors. I believe that, because of leaky vessels in these cases, extravasation of plasma proteins may lead to congelation and fibrosis so that in due time the initially *humid* tumor may transform itself into a quite fibrotic one.

Finally, some meningiomas for reasons unknown, perhaps because of some chemical substances or antigens they may possess, have a tendency of attracting large numbers of lymphocytes and plasma cells. This feature was first noted by BANNERJEE and BLACKWOOD (1971) who thought they were dealing with a mixed meningioma and plasmacytoma. However, the lymphoreticular elements are cytologically benign and their polyclonal character has been documented (STAM et al. 1980). While several cases of such tumors have found their way into the literature (HORTEN et al. 1979; MIRRA et al. 1983), for the most part they were regarded as only histologic curiosities (but with a potential for diagnostic mistakes, if only a small portion of such tumors, featuring only the inflammatory cells, were to be examined).

First in Taipei and then in other parts of the world, we have found some meningeal tumors with a particularly prominent myxoid/chordoid cell population, that may attract such large masses of plasma cells and lymphocytes, sometimes with large follicles and vascularized germinal centers, that the latter may behave as autochtonous growths and produce systemic effects identical to those seen in patients with so-called angiofollicular lymphoid hyperplasia in the Castleman's syndrome (KEPES et al. 1988). Such complications, that so far we have seen only in young patients, include stunted somatic and sexual development and above all iron-resistant hyperchromic microcytic anemia and hyperglobulinemia, all of which usually clear up beautifully after the offending meningioma and its lymphoplasmacellular infiltrate have been surgically removed. In cases of local recurrences of such tumors, the return of the anemia has also been observed.

The following is an outline that I feel might provide useful additions to the subclassifications of meningiomas in any new or revised edition of the WHO "blue book."

Proposed Changes in the Classification of "Tumours of Meningeal and Related Tissues" (pages 53-56 in the World Health Organization's *Histological Typing of Tumours of the Central Nervous System*, Geneva, 1979)

Comments regarding the present text of the WHO Classification: The present text defines *meningioma* as "a tumour originating from cellular elements of the meninges". I submit that we do not adhere to this concept in our everyday diagnostic practice and even the WHO classification departs from its own definition. Example: melanocytes are normal constituent cells of the leptomeninges, yet primary melanomas of the meninges are listed separately from *meningiomas*. There are also primary leptomeningeal malignant lymphomas and we call those *lymphomas* and not *meningiomas*.

It would probably be more useful to restrict the term *meningioma* only to tumors that can be shown to have their derivation from arachnoidal cap cells.

The term *meningioma* was coined by Cushing more than 65 years ago and it has served us well. It certainly represented progress compared to earlier terms borrowed from known tumor entities in general pathology, such as *Krebs, fibroblastoma, endothelioma, sarcoma angiolitique etc.*, which were just as inappropriate for this group of tumors as *melanocarcinoma, melanosarcoma* used to be for malignant melanomas. But while in other fields of pathology eventually *cell-related* definitions were given to previously hard-to-define tumors, as for example, mesotheliomas, Cushing named meningiomas after *topographically defined gross anatomical structures,* i.e. the meninges. This approach can be regarded as analogous to call pleural of pericardial mesotheliomas *pleuromas* and *pericardiomas* respectively, or to designate a malignant synovial tumor originating from a bursa as a *bursoma.*

The WHO classification adhered to Cushing's concept in some instances, as in the case of category 6 of meningiomas (page 55) called *hemangioblastic.* I submit that classical hemangioblastomas may involve the leptomeninges, a number of such cases are to be found in the literature and the cases best suited for analysis are from individuals who have the full-blown von Hippel-Lindau syndrome and happen to have simultaneously cerebellar, possibly brainstem and meningeal masses with identical histology of hemangioblastoma. In my opinion it is a mistake to call the meningeal representative of this tumor a *meningioma* when in other locations it is simply called a *hemangioblastoma.* They are not meningiomas and meningothelial cap cells are not part of their make-up. They have the characteristic stroma cells accompanying the capillaries and venules as in other locations and these cells do not make any attempt to wrap themselves around vessels as meningothelial cells do. I would place such typical hemangioblastomas into category E (Others) as presently printed on page 56, among fibromas, chondromas, etc.

As to hemangiopericytomatous tumors, whether or not they should be regarded as a form of meningiomas, again should depend on the consensus as to their relationship to other meningiomas. I happen to believe that meningeal hemangiopericytomas (hpct-s) are not just soft tissue hpct-s involving the leptomeninges. They are much more common there than the number of vessels in the meninges would support in a statistical sense (total weight of leptomeninges is only about 30 grams in a grownup weighing 75 kgs = 75.000 grams). Transitional forms to meningothelial meningiomas have been observed etc. At any rate, their inclusion among meningiomas should depend on their relationship to arachnoidal cap cells, otherwise they too should be listed in group E as *others.*

I would leave category 5 (*angiomatous*) intact and agree with its description.

I would propose to list #4: psammamatous variant not independently, but as a subtype of *meningotheliomatous* type, as Cushing and Eisenhardt did in their monograph because, if anything, whorl formation with or without secondary hyalinization and calcification is truly the most notable histological feature of *meningothelial* cells (arachnoidal cap cells) in their normal or neoplastic condition. Just how many psammoma bodies per field are required before a meningioma is called *psammomatous* has never been specified and, for this reason too, I suggest that psammomatous meningiomas should be part of the *meningotheliomatous* group.

Suggested Entities to be Considered by the Group, for Addition:

1. Meningiomas that contain large numbers of intracellular hyaline inclusions or pseudopsammoma bodies form an interesting subgroup for which ALGUACIL-GARCIA et al. (1986) suggested the name *secretory meningioma.* The morphological appearance of this variant is noteworthy, it highlights certain important *epithelial* characteristics of meningiomas (the ability of forming intracellular lumina, the switching from vimentin to keratin in its intermediate filaments, production of carcinoembryonic antigen, epithelial membrane antigen, secretory compound, etc.) and they also have a much higher incidence of perivascular proliferation of pericytes than ordinary meningothelial meningiomas. The morphological make-up of these tumors also has some possible clinical connotations, one being the fact that almost without exception hyaline inclusions are seen in benign meningiomas, but also it appears that they may be more often accompanied by massive peritumoral edema than other meningiomas of similar locations or size.

2. It has been known for quite some time that certain meningiomas (probably by virtue of some heretofore undefined antigen) attract lymphocytic-plasmacellular infiltrates which at times may be so pronounced as to overshadow the meningiomatous component of the intracranial mass. At times, particularly in the case of meningiomas with myxoid/chordoid alterations, the lympho-plasmacellular infiltrate can be so heavy, complete with follicles and germinal centers, that systemic complications may develop. These include (particularly in young people) retarded somatic development, iron-resistant microcytic hypochromic anemia, hyperglobulinemia and sometimes hepatosplenomegaly, resembling the Castleman syndrome, which encompases systemic complications of *angiofollicular lymphoid hyperplasia or hamartoma* lymphoplasmacellular mass seen in the mediastinum and other locations. In the case of such meningiomas, apparently the lympho-plasmacellular cell mass, although probably only reactive in character, may produce the same systemic clinical alterations as the genuine Castleman tumor does.

   For these reasons, I believe a subgroup designated as *Meningiomas with massive lympho-plasmacellu-*

lar infiltrates would be a justifiable addition to the existing categories.

3. *Meningio-angiomatosis*, a probably hamartomatous condition of the leptomeninges and cerebral cortex could be considered for inclusion as an entity.

Question: Should the photomicrographs of the booklet and the slide collection include only the *typical* presentations of the listed categories, or should less common, but by no means rare, microscopic features be included, particularly those that may lead to the mistaken diagnosis of another type of neoplasm? Alternately, should such details be left to past and future textbooks, atlases and monographs on brain tumor pathology? (Examples: Masson's *forme humide, meningiomas with Verocay-like bodies, imitations of gemistocytic astrocytomas, etc.)*

CHAPTER XVII

Biologic Behavior of Meningiomas

KURT L. JELLINGER, *Boltzmann Institute of Clinical Neurobiology, Lainz Hospital, Vienna, Austria*

Most meningiomas are well-encapsulated slowly growing tumors that are amenable to total resection and are therefore considered benign. Until recently, the extent of the surgical removal of these tumors was considered as one of the most important prognostic factors. Recurrence rates of meningiomas, following total resection, are estimated to range from 5 to 30% with much higher figures after subtotal removal. The average recurrence-free intervals range from 5 to 19 years. In a large series (JELLINGER and SLOWIK 1975), the recurrence rate in patients surviving at least 5 years after total resection was 13%, while a recent study from the Massachusetts General Hospital (MIRAMANOFF et al. 1985) calculated a risk of recurrence after total resection of 7% at 5 years, 20% at 10 years, and 32% at 15 years.

The biologic behavior of meningiomas is often not predictable from the histologic appearance of these tumors alone, and ultrastructural or tissue culture studies have not been conclusive in elucidating this problem. It is generally accepted that no obvious relationship exists between the histologic subtype and recurrence rate in meningiomas, except for the established intermediate and anaplastic types (i.e. hemangiopericytoma, papillary and anaplastic meningiomas). The incidence of these malignant forms ranges from 0 to 15.8%, and in our series was 3.3 and 3.7%, respectively. It is generally agreed that these forms show a significantly higher recurrence rate with or without a shorter progression rate and a tendency to metastasize outside the CNS. However, no definite relationship with a tendency to recur has been found for the other subtypes of *classical* meningiomas (Table I).

With regard to the prognostic significance of the histologic features in meningiomas (Table II), most authors agree that hypercellularity, cellular pleomorphism, diffuse growth, high mitotic rate, tumor necroses, cortical and/or bone invasion correlate with aggressive behavior, a faster growth rate and more frequent recurrence. On the other hand, several authors found little correlation between the histologic characteristics and the growth rate of these tumors (SIMPSON 1957). These findings are supported by the occasional

TABLE I Histologic subtypes (%) in subtotally resected meningiomas with and without recurrence

Histologic Subtype	Crompton & Gautier (1970)		Skullerud & Loken (1974)		Jellinger & Slowik (1975)		de la Monte[a] et al. (1986)		Pers. ser. (1977-87) Survival > = 5 years	
	Recurr	Non-R	Recurr	Non-R	Recurr	Non-R	Recurr	Non-R	Recurr	Non-R
Meningothel.	35	16.7	50	53	55	68	68	57	74	68
Transitional	60	52.5	29	31	15	21	47	50	12	17
Fibroblastic	5	25.7	8	12	9	8	58	77	0	3
Psammomatous	--	--	--	--	0	0	16	39	0	8
Angiomatous	--	--	--	--	0	0	--	--	0	1
Angioblastoma	--	--	--	3.6	0	2	--	--	0	0.5
Pericytoma	--	--	12.5[x]	0.7	12[x]	0.5	11[x]	2	8[x]	1
Anaplastic	--	--	0	0	9[x]	0.5	21[x]	0	6[x]	1
TOTAL NO. CASES	20	78	24	137	34	236	38	44	74	381

(a) Percentage exceeds 100% due to overlap of subtypes; (x) $p<0.001$

TABLE II Histologic features in subtotally resected meningiomas (2 series)

	Skullerud & Lokena[a] (1974)		Jellinger & Slowik (1975)	
Histologic Type	Recurr %	No Recurr %	Recurr %	No Recurr %
Hypercellularity	25[x]	1	46[x]	11
Hypervascularity	8	4	15	14
Diffuse Growth	33[x]	1		
Nucl. Pleomorphism	8	13	11[x]	3
Increased Mitotic R.	8	0	38[xx]	3
Focal Necroses	17	13	26	10
Bone Invasion	25[x]	6	15[x]	3
Cortex Invasion	0	1	18[x]	5
TOTAL NO. CASES	12	72	26	185

(a) Only syncytial type
(x) $p<0.05$
(xx) $p<0.001$

observation of mitoses, necrosis, and even cortical invasion in meningiomas that do not recur after surgery. CHRISTENSEN et al. (1983) demonstrated bone invasion in 3.8% of the meningiomas without recurrence and in 15% of recurrent tumors ($p<0.02$), while cortical invasion was not significantly different in both groups. Only necrosis was found in 26% of recurrent tumors as compared with 3% in non-recurrent ones ($p<0.01$). The nuclear counts were identical in paraffin sections from tumors with (6.4 nuclei/1000/mm^2) and without recurrence (6.1/1000/mm^2), while the cell count in the recurrences was higher (7.4 nuclei/1000/mm^2) than in the primary tumors. No differences were found for the nuclear area fractions between recurrent and non-recurrent meningiomas, while the nuclear area fraction was increased in the recurrences.

However, not all meningiomas that recur have any of the above histopathologic features. JÄÄSKELAINEN et al. (1985), in a study of 43 patients with meningiomas, used hypercellularity, loss of architectonic pattern, nuclear pleomorphism, mitotic index, focal necrosis and brain infiltration to grade meningiomas as follows:

I - benign, II - atypical, III - anaplastic, IV - sarcomatous. These authors concluded that the mitotic index correlated with the growth rate, but its value in predicting tumor doubling time is only suggestive and the findings regarding cytologic pleomorphism are difficult to correlate with tumor recurrence or non-recurrence (KEPES 1982). Moreover, the mitotic rate of meningiomas is too low to be properly measured and the distribution of mitotic figures is too inconsistent to provide the basis for either a reliable estimate of proliferation potential or prediction of the growth rate of individual tumors. In a series of 82 subtotally resected meningiomas with 44 recurrences, DE LA MONTE et al. (1986) found 7 among 20 semiquantitatively evaluated features to be positively correlated with tumor recurrence: increased vascularity, hemosiderin deposits, growth of tumor cells in sheets rather than the usual pattern, prominent nucleoli, mitotic figures, single-cell and small-group necroses, nuclear pleomorphism, and overall atypical or malignant tumor grade ($p<0.005$). In addition, sheeting of tumor cells, prominent nucleoli, and the presence of a less than 10 % meningothelial pattern were correlated with significantly shortened progression-free intervals between surgery and tumor recurrence.

A semiquantitative evaluation of the histologic features in 445 subtotally resected meningiomas (1977 to 1983), 68 of which underwent a second or further surgery for recurrence at least 5 or more years after first operation, gave similar results histologic features showing positive correlation with recurrence were: sheeting of tumor cells, hypercellularity, mitotic figures, nuclear pleomorphism, cortical and bone invasion, and atypical or malignant tumor grade ($p<0.001$), while less significant correlations were found for hypervascularity, micronecrosis, and prominent nucleoli (Table III).

These findings suggest that the presence of some features, such as prominent nucleoli, growth of tumor cells in sheets, possibly indicating a lower order of differentiation compared with whorls or syncytia, individual cell and small-group necroses, and nuclear pleomorphism, may be used as additional criteria to predict possible recurrence of totally resected meningiomas that would not be classified as malignant by traditional criteria.

Similar problems arise in the evaluation of recurrent tumors. While most recurrent meningiomas do not change their basic histopathologic pattern, in our previous series of 88 totally removed "classical" meningiomas, eleven (12.5%), showed some histologic signs of increased aggressiveness in later biopsies. In 7 cases, the first biopsy specimen showed the *silent* picture of various types of meningiomas without mitotic figures, in four other specimens high cellularity and few mitotic figures were present in the first biopsy. In 9 patients, the tumor removed at a second or further operations, displayed increased cellularity and mitotic rate and/or focal necrosis without having changed its basic architecture. Only 2 cases later developed a fibrosarcoma.

TABLE III Histologic features of 445 (sub)totally resected meningiomas (pers. series, survival >= 5 years)

Histologic Feature	Recurrent N (n=68)	%	Non-Recurrent N (n=377)	%
Whorls	52	76.5	210	55.7
Psammoma bodies	24	35.4	231	61.3
Hyaline inclusions	6	8.8	37	9.8
Intranucl. vacuoles	17	25.0	68	17.0
Lymphocyt. infiltrates	3	4.4	9	2.4
Metaplasia	6	8.8	37	9.8
Xanthomatous foci	3	4.4	14	3.7
Hypervascularity	30	44.1(x)	107	28.4
Hemosiderin deposits	8	11.8	17	4.5
Sheeting	40	58.8(xx)	32	8.5
Nucleoli	7	10.3(x)	10	2.6
Mitotic figures	17	25.0(xx)	27	7.2
Micronecroses	10	14.7(x)	20	5.2
Pleomorphism	10	14.7(xx)	12	3.2
Hypercellularity	14	20.6(xx)	10	2.6
Brain invasion	2	3.0(xx)	1	0.3
Bone invasion	7	10.3(xx)	6	1.6
Giant cells	2	3.0	2	0.5
Tumor grade II or III	12	17.6(xx)	4	1.1

(x) $p<0.05$
(xx) $p<0.001$

In a recent sample of 74 recurrent meningiomas, some showing up to 5 or 7 operations, there were only three cases developing histopathologic signs of malignancy: An orbital tumor remained unchanged up to the sixth recurrence 12 years after first surgery, while the seventh recurrence, one year later, showed anaplasia with greatly increased mitotic rate, pleomorphism, and invasive growth. Two other cases of meningothelial meningiomas, at second recurrence, 5 and 8 years after first surgery, showed increased cellularity, pleomorphism and mitotic figures without definite features of sarcoma. Although histopathologic techniques can demonstrate invasiveness and anaplasia, the actual proliferating potential of meningiomas cannot be determined conclusively from its histology.

TABLE IV Labeling index of BrdU in meningiomas (first operation)

Histologic Subtype			Labeling Index %
Meningothelial	A	n = 35	Median <1.0 (>1;n=6)
	B	n = 14	SD 0.8 (0-1.6)
		n = 1	2.3
	C	n = 1	2.0
Fibroblastic	B	n = 1	0.1 (0-0.4)
	C	n = 1	0
Transitional	B	n = 9	SD 0.3 (<0.1-6.3)
	C	n = 2	SD 0.25 (0.1-0.4)
Angiomatous	B	n = 2	SD 0.1 (0-0.4)
Anaplastic	A	n = 5	2.8 ± 0.8 (<1-3.8)
	A	n = 5 (Recurrent)	4.5 ± 0.8 (2.9-5.4)
	C	n = 1	2.0

A CHO et al. (1986, 1987)
B IWAKI et al. (1987)
C MORIMURA et al. (1989)

Useful information on the biologic behavior of meningiomas has been provided by recent cytokinetic studies using monoclonal antibodies against proliferation markers, e.g. bromodeoxyuridine (BrdU) or Ki-67. The average labeling index (LI) for BrdU in non-malignant meningiomas is less than 1%. This was seen in 27 (82%) of 33 non-malignant meningiomas, but was 3.6% in 10 malignant tumors in the series of CHO et al (1987), five of which were recurrent tumors. All of the malignant meningiomas had LI greater than 2.7% except for one that was diagnosed solely on the basis of invasiveness. In Cho's series, nine (56%) of 16 meningiomas with an LI greater than 1% were recurrent, whereas only 3 (10%) of 29 tumors with LI less than 1% recurred. Thus, meningiomas with LI greater than 1% appear to grow faster and recur more frequently than those with LI less than 1%. The tumor doubling time estimated from serial computed tomography (CT) scans in 8 patients with recurrent meningiomas ranged from 8 to 440 days and showed close inverse correlation with the BrdU LIs (correlation coefficient of 0.99) (Table IV).

In two series of recurrent meningiomas, most recurrent tumors did not show an increased LI at second or further operation, in one case even at the 7th recurrence (CHO et al. 1987; IWAKI et al. 1987). Two cases, however, recurring after short periods, at the third and fourth operations showed high LI of 9 and 13.6%, respectively, associated with increased mitotic rate and necrotic foci. Increased LIs were also seen in recurrent pericytomas and malignant meningiomas. Thus, there is a definite correlation between high LI and rapid recurrence (IWAKI et al. 1987) (Table V).

TABLE V Labeling index (BrdU) in recurrent meningiomas

First Recurrence			
Meningothelial	A	n = 5	<1.0 (3 >1)
	B	n = 4	0.4 (0-3.6)
Transitional	B	n = 2	0.5 (0.2-1)
	B	n = 1	13.6 (6.3-28)
Angiomatous	B	n = 1	0.3 (0-0.5)
Malignant	A	n = 5	4.5 ± 0-8
Pericytoma	A	n = 2	2.3 (0.5;4.1)
Second Recurrence			
Meningothelial	B	n = 1	0.5 (0.2-1)
Anaplastic	B	n = 1	9.0 (0.2-17.5)
Seventh Recurrence			
Meningothelial	B	n = 1	0.3 (0.2-0.5)

A CHO et al. 1987
B IWAKI et al. 1987

Estimation of the growth factor determined by Ki-67 immunoreactivity has provided a wide range of results and inhomogeneous distribution of proliferating cells in various subtypes of meningiomas. In meningothelial and fibroblastic tumors, the Ki-67 LI ranges from 0.3 to 3.2% (Table VI), although ROGGENDORF et al (1987) saw a wider range up to 7.8%, and increased LI in transitional type tumors (Table VII). Anaplastic meningiomas show proliferation rates of 9 to 20% (mean 15%), and one pericytoma had an LI of 31%, with diffuse distribution of labeled cells. In most recurrent tumors, an increased Ki-67 LI has been found in both *classical* and anaplastic meningiomas (Table VI and VII).

Comparison of LIs obtained with Ki-67 and a monoclonal antibody to BrdU showed no significant differences in isomorphic and anaplastic gliomas, and meningiomas, whereas the Ki-67 LI in metastatic carcinomas was higher than that of BrdU (Morimura et al. 1989). These data are in agreement with those by ROGGENDORF et al. (1988) who also reported higher Ki-67 indices than are commonly found for BrdU. Some of these discrepancies may be due to technical factors of different counting procedures and, in particular, by a significant heterogeneity of proliferation patterns occurring within one sample from area to area, as well as between different samples of the same tumor. According to recent data, however, the *in vitro* labeling studies with BrdU allowing identification of S phase cells compares favorably with Ki-67 labeling which identifies all proliferating cells, i.e. the growth fraction (MORIMURA et al. 1989). Further cytokinetic studies obtained by both BrdU and Ki-67 labeling methods may increase the accuracy with which the proliferation potential of meningiomas can be predicted. By supplementing the histologic diagnosis performed with routine methods, these studies may help to improve the determination of prognosis. However, further correlative studies of histopathology, cytokinetic, and clinical data are needed.

TABLE VI Proliferation index (Ki-67 labeling) in meningiomas (first operation)

Meningothelial	A	n = 4	0.5 ± 0.3
	B	n = 11	SD 1.8 (0.2-4.0)
	C	n = 4	3.2 ± 2.5 (1-6)
	D	n = 9	SD 2.5 (<1.0-6.0)
	E	n = 10	SD 1.1 (0-3.8)
	F	n = 2	SD 1.7 (0.7-3.0)
Fibroblastic	D	n = 2	1.0 (<1.0-1.9)
	E	n = 1	0.9
Transitional	D	n = 5	7.5 (1.9-12.3)
Psammamatous	C	n = 1	0
	D	n = 2	9.7 (9.5,10.0)
Anaplastic	D	n = 2	14.9 (9.3,20.5)
	E	n = 1 (recurrent)	10.0
	F	n = 1	3.0
Pericytoma	D	n = 1	31.0

A = BURGER et al. (1986); B = KLEIHUES et al. (1987); C = GIANGASPARO et al. (1987); D = ROGGENDORF et al. (1987); E = REIFENBURGER et al. (1989); F = MORIMURA et al. (1989)

* * * * * *

TABLE VII Proliferation rate (Ki-67) in meningiomas (ROGGENDORF et al. 1987, 1988)

Meningothelial	n = 9	2.5 (<1.0-7.7)
Mening/infiltr.	n = 1	10.9
Recurrent	n = 1	8.5
Fibromatous	n = 2	1.0 (<1.0-1.9)
	n = 1	15.4
Recurrent	n = 1	6.6
Angiomatous	n = 3	1.3 (1.1-1.5)
Transitional	n = 5	7.5 (1.9-12.3)
Psammamatous	n = 2	9.7 (9.5,10.0)
Anaplastic	n = 2	12.5 (12.1,12.9)
Recurrent	n = 2	14.9 (9.3-20.5)
Pericytoma (recurrent)	n = 1	32.4

CHAPTER XVIII

Meningiomas: Immunocytochemistry and Steroid Hormone Receptors

JANET M. BRUNER, *Department of Pathology, The University of Texas M.D. Anderson Cancer Center, Houston, TX*

Introduction

This paper will serve as a review of published studies commenting on the immunocytochemical findings and steroid hormone receptor assays performed on several series of intracranial meningiomas.

Immunocytochemistry

Meningiomas have been studied with antibodies to several groups of antigens, usually with an intent to clarify histogenetic origin of the tumor, or to aid in differential diagnosis of difficult or unusual cases (ALGUACIL-GARCIA et al. 1986; THEAKER et al. 1986; MEIS et al. 1986; HALLIDAY et al. 1987; HITCHCOCK and MORRIS 1987; HOLDEN et al. 1987; MANIVEL et al 1988, WINEK et al. 1988; ZARBO et al. 1988). Antibodies to intermediate filaments, epithelial antigens or neuroectodermal markers have all been used. Data from several published and presented series are summarized in Table I.

The conclusion of most authors is that meningiomas almost universally contain diffuse or membranous cytoplasmic epithelial membrane antigen. Despite this, cytokeratins are not so prevalent, and are usually associated with the eosinophilic bodies found in the variety known as *secretory* meningioma. Vimentin is usually strongly and uniformly present. S-100 protein, considered a neuroectodermal antigen, is present in a highly variable number of cases. Universally present when sought are fibronectin and desmoplakin (HITCHCOCK and MORRIS 1987), although these antibodies have not been used by many authors, since they may require special handling of tissue. Antigens which are usually not found include glial fibrillary acidic protein, desmin and Leu-7.

An interesting comparison with these same antibodies has been made by some authors in sections of normal arachnoid or arachnoid granulations. This data has shown nearly universal positivity for epithelial membrane antigen and vimentin, as with meningiomas. S-100 protein has been absent. Cytokeratin has shown variable staining.

TABLE I Immunocytochemistry of meningiomas

	#Cases	EMA	CK	VIM	S-100
HALLIDAY 1985	16	NT	0	16	NT
MEIS 1986	50	25	12	9	4
THEAKER 1986	3	13	5	13	13
HOLDEN 1987	14	NT	7	14	NT
HITCHCOCK 1987	50	NT	28	50	39
WINEK 1988	41	41	5	41	6
		175/200	92/280	238/280	85/250

(NT = Not Tested)

Despite these rather exhaustive investigations, no firm conclusions can be drawn from them as to the embryonic origin of the meninges or of meningiomas. These tumors still seem to have some features of both epithelial and neuroectodermal cells. However, it can be re-emphasized that the differential diagnosis between meningiomas and glial neoplasms may be assisted by markers for epithelial membrane antigen and glial fibrillary acidic protein. Another not insignificant problem, differentiation between atypical meningiomas and metastatic carcinoma, is not possible to define by using immunocytochemical markers. A variable number of each type of tumor may show positivity for cytokeratin and epithelial membrane antigen.

A comparison which has been only occasionally investigated is that between angioblastic or hemangiopericytic meningiomas and hemangiopericytomas of soft tissues (HOLDEN et al. 1987; WINEK et al. 1988). One group found analogous staining between normal arachnoidal cells, meningiomas, and hemangiopericytomas (HP) of the meninges (HOLDEN et al.1987), and concluded that HP of meninges were thus a variant of meningioma. The other authors (WINEK et al. 1988) compared soft tissue HP with meningiomas and HP of meninges. Staining of meningiomas was distinctly different from that of both types of HP, which showed a similar pattern with vimentin, cytokeratins, and epithelial membrane antigen. He concluded that HP of meninges were probably a distinct tumor type, and not a variant of meningioma. More studies undoubtedly need to be performed before either conclusion is universally accepted.

Steroid Hormone Receptors

The presence of steroid hormone binding proteins in meningiomas has been investigated because meningiomas have a higher incidence in women than in men, particularly in women between ages 25 and 60 years. Meningiomas also have an accelerated growth rate during pregnancy, and there is a more than random association of the occurrence of meningioma in women with breast carcinoma (HINTON et al. 1983; PERTUISET et al. 1984).

Figure 1 shows the theoretical mechanism of estrogen receptor binding and stimulation of progesterone receptors. Estrogen enters a cell, binds to a steroid hormone (estrogen) receptor, and is translated to the nucleus. The presence of the hormone-receptor complex stimulates protein synthesis. One of the proteins that is synthesized in response is the progesterone receptor (PR). Thus PR activity is classically considered a biochemical marker of estrogen action (MARKWALDER et al 1984). In most recent studies which have examined estrogen receptors (ER), PRs have also been investigated (HINTON et al. 1983; MAGDELENAT et al. 1982; POISSON 1984, HAYWARD et al. 1984; BLANKENSTEIN et al. 1984; MARKWALDER et al. 1984). Both types of receptors have been sought in the cytoplasm and the nucleus of meningioma cells.

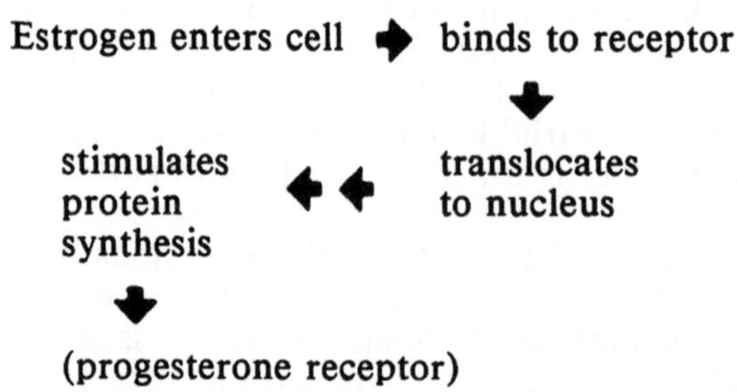

Figure 1. Estrogen and Progesterone in Meningiomas

The results of these studies have been conflicting, depending on the quantitative level of steroid hormone receptor considered as positive or negative (HAYWARD et al. 1984; BLANKENSTEIN et al. 1984). An unexplained finding is the apparent discrepancy between levels of ER and PR positivity, both qualitative and quantitative. There are fewer cases in which ER positivity is shown, and the levels are generally much lower than for PR (SCHNEGG et al. 1981; COURIERRE et al. 1985; POISSON et al. 1983; MOGUILEWSKY et al. 1984). This might be explained by the level of receptor binding, the presence of different receptor epitopes, or the activity of the hormones or the receptors. It might also be due to types of quantitative methods used to elucidate the presence of receptors. Immunocytochemical detection of these receptors might help in answering some of these questions, and the method is just coming into use for paraffin tissue sections (CHENG et al. 1988). Previous studies using frozen tissue sections (HINTON et al. 1983) found cytoplasmic and nuclear labelling for ER in 4 of 11 cases, and cytoplasmic staining only for PR in 5 of 11 cases. It will be interesting to discover if there is some predictive or therapeutic value in the measurement of ER and PR in meningiomas, as there is for breast carcinoma. To date there has been no significant correlation of ER/PR status with clinical or histopathologic features of meningiomas (MAGDELENAT et al.1982).

Since manipulation of tumor growth rates is theoretically possible using estrogen and progesterone analogs, several *in vitro* studies using cell culture manipulation have been performed (Figure 2). Compounds tested have included Tamoxifen (an estrogen antagonist), estradiol, estrogen and progesterone (JAY 1985; OLSON et al. 1986). The presence of estrogen and progesterone agonists, including estradiol, increase the *in vitro* growth rate of cultured meningioma cells. Tamoxifen has shown a variable effect on growth rate, either increasing or decreasing, apparently dependent on the length of time the meningioma cells are kept in tissue culture. Investigation of this drug is continuing. Another estrogen antagonist, Nafoxidine, and a progesterone antagonist, RU38486, both inhibited the *in vitro* growth of meningioma cells (OLSON et al. 1986).

Estradiol	↑ or no effect
Progesterone	↑ or no effect
Tamoxifen	↑ or ↓
Androgen (DHT)	no effect

Figure 2 In vitro growth manipulation

Recent studies using androgen agonists for *in vitro* growth manipulation have found no measurable effect on growth rate (Olson 88). This despite the finding of androgen receptors in meningiomas (SCHNEGG et al 1981; POISSON et al 1983; PERTUISET et al 1984; MOGUILEWSKY et al 1984).

Current data do not permit conclusions correlating *in vitro* responses or the presence and significance of ER/PR with clinical behavior or *in vitro* growth modulation of meningiomas. However, some promise in this area justifies continuing interest.

General Discussion of Chapters XVI–XVIII

DR. EARLE: We may be premature in changing the meningioma classification for the same reasons which are applicable to the other changes that have been proposed. Historically, when I first went to the Armed Forces Institute of Pathology in 1962, I was giving consultations on consultations. Dick Davis and I were confronted with the fact that Webb Haymaker and John Waggener had quite consistently diagnosed meningiomas according to the Cushing and Eisenhart classification. They would send out diagnoses such as, "This is a meningioma type 6, variant 4," which was absolutely devastating to the pathologist who received the consultation, unless he happened to have the book. So we found ourselves having to write a consultation explaining what the consultation was about. Then, when Kernohan's book became popular, we had only three types of meningioma: meningothelial, fibroblastic, and mixed. That was probably an oversimplification because, as John Kepes showed today, there are a lot of variants, although some are quite rare.

The term *meningioma*, if you recall, was proposed to differentiate between the more devastating glial tumors and the more benign tumors of the coverings (meninges) of the nervous system. It seemed illogical to us, when we were discussing angioblastic meningiomas during sessions for the first Blue Book, to eliminate the blood vessels from the coverings of the brain. However, I think that certain terms we used in the WHO Blue Book have not received as much acceptance as we intended. I have no quarrel at all with eliminating the hemangiopericytoma and hemangioblastoma from the meningioma category.

However, isn't this somewhat in contrast to what we have been saying? The prognosis of meningiomas has always troubled me and I thought it was very interesting to hear from Dr. Jellinger that pleomorphism of nuclei in meningiomas has never predisposed to a malignant behavior in the absence of mitoses or evidence of invasion. I think that needs to be brought out in our classification.

DR. SCHEITHAUER: With meningiomas as well as with other CNS tumors, if we are going to change the classification, it ought to be either to illustrate things that are visually very distinct, are likely to be confused with other things, or are lesions which have special clinical significance. I doubt that anyone is going to disagree.

As far as the angioblastic meningiomas or hemangiopericytomas (whichever term you care to use) are concerned, I agree with Dr. Earle. Perhaps not in all instances, but in most, one can distinguish this lesion from meningioma. One problem is that we define the entity by H & E stain, then proceed to study it by more sophisticated methods, and come to the conclusion that the tumors under study are a mixed group that cannot be distinguished with certainty.

Perhaps we should look at them with different techniques, particularly with immunocytchemistry. Having done that, I am convinced that the meningiomas and hemangiopericytomas are, in most instances, two distinct entities. There are occasions, however, when we will have difficulty in making he distinction. Sometimes it is difficult to separate them because of poor fixation, a non-representative biopsy, or other factors, but in most instances, I feel that I am able to distinguish between the two.

With reference to a remark Dr. Kepes made earlier, in looking at about 40 angioblastic meningiomas, I was struck by the fact that only in one instance were we able to find any papilla formation. I do not believe that so-called *papillary meningiomas* take their origin from the meninges to any significant extent.

I might mention that in our immunochemical studies, hemangiopericytomas were all epithelial membrane antigen (EMA)-negative, whereas all meningiomas were EMA-positive. Although that is only one marker, it does suggest that there is a dividing line between these two lesions.

With regard to other lesions that Dr. Kepes brought to our attention which presently are not in the WHO Blue Book, the secretory meningioma should be recognized, though it is not a morphologically distinct or clear clinicopathologic entity. Pseudosammoma bodies are not necessarily found only in that tumor. One can find them in so-called *microcystic meningiomas* as well. John Kepes showed a very spongy appearing type of meningioma that often is associated with striking edema on CT scan. For the clinician, this lesion may constitute a meaningful subcategory of meningioma.

I am ambivalent about the chordoid meningioma because I have not had the experience that Dr. Kepes has had. The few I have seen have been EMA-negative. That makes me a little uncomfortable in calling them *meningiomas*.

I recognize the plasma cell-rich meningioma. Dr. Richard Davis has shown me examples that were incontrovertibly meningiomas with plasma cells. I have also seen cases wherein the plasma cells and the reactive infiltrate were so abundant that we really did not know if it was a plasma cell granuloma. Another question is, "To what degree can we ascertain that the plasma cells are reactive rather than neoplastic elements of the process?" I think the pathologic spectrum of plasma cell-rich lumps in the dura is broader than we have been led to believe.

In my opinion, the hemangioblastoma is a distinct lesion that has no place in the spectrum of meningiomas.

DR. RUBINSTEIN: First, in answer to Ken Earle, I do not think that the majority of hemangiopericytic tumors arise from the dura. Many of them do in fact arise from the arachnoid, and many of them seem to be within the parenchyma of the brain.

Secondly, I think that John Kepes has made a very convincing case for separating the meningeal arachnoidal cells from any other cell in the body and also for separating the hemangiopericytic tumors, or the angioblastic meningiomas from the other hemangiopericytomas. I think his argument is very convincing.

Thirdly, every neurosurgeon and every pathologist knows that the angioblastic meningioma, pericytic type, is a more malignant tumor, which will recur earlier. Whether you call it hemangiopericytoma or angioblastic meningioma makes no difference. Although the histogenetic argument is convincing, disagreement remains regarding the significance of some of the histochemistry. Since there is dispute, perhaps we should leave things the way they are in the Blue Book. We should maintain the decision arrived at previously, to have the angiomatous, angioblastic, hemangioblastic, hemangiopericytic variants in the Blue Book as stated. Also, having seen several cases, I think that there is a very frequent association between the papillary pattern and the hemangiopericytic type of meningioma.

The spaper published 18 years ago (PITKETHLY et al. 1970) made the distinction between the hemangioblastic and the hemangiopericytic variant crucial. It is a very important contribution and is a landmark in providing clarification.

I would like to say something about the mysterious association of lymphocytic and plasmacytic elements within meningiomas and add that we have seen recently a very remarkable example of a patient in whom a meningioma recurred three times, although at two different sites, so there were two meningiomas. The first was a meningioma with a classical meningothelial pattern. The second was a little remote from the first, but distinct. It was also a meningioma but with a striking lymphoplasmacytic infiltration (BANERJEE and BLACKWOOD 1971; HORTEN et al. 1979). Later, a recurrence of the second tumor took place. This time the plasma cell infiltration was so striking that one of our residents decided to test for the presence of monoclonality, using the light and heavy chain markers. It turned out to be a monoclonal expression of the plasma cell. This patient was diagnosed on the basis of monoclonality as having had the meningioma transform into a lymphoma. I believe that we are going to see a transformation of these polyclonal infiltrations into a monoclonal pattern. Then we will have to draw a conclusion as to what the transformation signifies.

DR. ISHIDA: I would like to make two comments about meningiomas. The first relates to hemangioblastic meningioma. The WHO histologic classification describes three types of highly vascularized meningioma: angiomatous, hemangioblastic and hemangiopericytic. The angiomatous meningioma contains many large and small vascular channels whose walls appear hyalinized. Intervening clusters of cells show xanthomatous change. By electron microscopy the cells situated outside the thick layer of vascular basement membrane and collagen show features of meningotheliomatous cells as evidenced by intracytoplasmic filaments and intercellular junctions.

The WHO classification defines the hemangioblastoma type as a meningioma that is in most parts indistinguishable histologically from the capillary hemangioblastoma of the cerebellum, except by its location and its more distinct encapsulation. The hemangioblastic meningioma is rich in capillaries with intervening vacuolated cells. In electron microscopy the structure of the vascular channels is mostly that of capillaries. The cells seen outside the vessel show prominent desmosomal connections. Abundant intercellular spaces are seen as well.

Dr. Kepes mentioned that true hemangioblastomas do exist in the supratentorial region, but they might be rare. Even if we keep the category of hemangioblastic meningioma, most of the tumors designated as such may virtually be of a type (capillary type) of angiomatous meningioma.

The second point I would like to mention is the presence of a meningioma which contains large numbers of eosinophilic intracytoplasmic inclusions. The histology of this meningioma is that of transitional meningioma. Many tumor cells contain globular eosinophilic intracytoplasmic inclusions of varying size. Some are vacuolated. In contrast to pseudopsammoma bodies they are PAS negative. By electron microscopy the inclusions appear as osmiophilic bodies, the larger ones containing lucent areas. Intracytoplasmic filaments appear in intimate connection with the inclusions. These inclusions were first described by Goldman and his coworker in 1980. Although rare, I think this type of inclusion should be added in the description of meningotheliomatous meningioma.

DR. DAVIS: As many of you know, I have had the opportunity of working intimately with Takao HOSHINO and his coworkers over the years, and have learned a great deal from him. One of the things I learned from the results of labeling index studies is the inaccuracy of making the diagnosis of meningioma by examination of a few high power fields after a low power survey, and flipping through the slides rather quickly, and I fear rather carelessly. Prodded by him and the results of the labeling index, I began examining meningiomas only in the morning and not on Friday, and looking at all fields of the tumor with medium to high power. I have discovered, to my chagrin, that I have clearly been missing important things in at least some of these tumors.

I have now seen small foci with small cells or small foci of necrosis, often associated with multiple mitotic figures in one high power field, scattered throughout these tumors. We began identifying these after hearing the results of the labeling index. Now, I am glad to say, we are starting to predict the increased labeling index on the basis of these morphologic features, which I just had passed over in the past. Now I insist that the residents submit all pieces of all meningiomas for microscopic examination so at least we have a good sample of everything that is removed at operation. I look at them when I am fresh and willing to take the time and care to go over them histologically.

DR. KEPES: This is a very important point. Although we tried to differentiate between benign and malignant meningiomas, it is very important to remember, as Dr. Davis has pointed out, that there are a number of meningiomas that are focally atypical and malignant, rather than diffusely so (MICHAUD and GAGNE, 1983).

DR. WECHSLER: I think the WHO classification of the meningioma group is fairly good. However, I have some problems with it. I also would like to discuss whether or not the psammamomatous type of meningioma should be included as a separate entity among the subtypes. We do not make this diagnosis very often.

Then, as far as regressive changes are concerned, I would like to ask Dr. Kepes whether some meningiomas present not only microcystic, but also macrocystic degenerations. This was discussed in a recent publication (SOBEL and MICHAUD, 1985), and we have also addressed this issue. If you refer to the old book of CUSHING and EISENHARDT (1930), there were meningiomas with microcysts and very litle tumor tissue left. Both our group and SOBEL found that in these special cases, the hyaline inclusions contained serum protein and immunoglobulins.

I was fascinated by Dr. Kepes suggestion that we think of a phenomenon of different types of secretion and permeability changes within these tumors. However, before we include a separate type of secretory meningioma, I think we will need further investigation.

My third point is related to this little group in the Blue Book which is called *malignant or anaplastic meningiomas*. As far as different degrees of malignancy of meningiomas is concerned, we should consider subtypes. With the help of BrdU and Ki-67 techniques, we now have a better basis for grading.

My fourth point is addressed to the demonstration by Dr. Bruner of additional information which can be obtained by immunohistochemistry. I remember one case of a tumor located at the base of the skull where positive immunohistochemistry for vimentin was helpful in concluding that this was a small meningioma rather than a glioma.

Lastly, I would like to raise a question for Dr. Kepes regarding the differential diagnosis between benign and malignant fibrous histiocytomas.

DR. KEPES: To answer Professor Wechsler, I think malignant and benign fibrous histiocytomas involving the meninges should be listed under *E* in the Blue Book as *Others*. I do not think they are truly meningiomas.

As for the matter of edema and cysts, the microcystic meningiomas belong to two major categories. In one category there is separation of the tumor cells, with fluid deposited between them, producing microcysts or macrocysts. This differs from the other kind of cystic meningioma which the surgeon sees on x-ray. Although this looks like a lot of fluid within the tumor, it turns out to be a lot of edema around the tumor.

Although secretory meningiomas can create an undue amount of edema around them, that is not the main reason I suggested this as a category. It is a distinct morphologic variant and, even without further research, I think deserves to be listed as such.

DR. SCHEITHAUER: Just an additional comment about the psammomatous meningioma. We need to illustrate this lesion, but gracing it as a full-blown subtype of meningioma is not correct. Pseudopsammoma bodies are a cell product of meningothelial cells that sometimes appear sufficiently numerous to characterize this tumor. It simply needs to be illustrated.

The same thing can be said for angiomatous meningiomas. Meningiomas can be highly vascular, but does this matter? The really unfortunate aspect of this term is that it introduces into the classification scheme yet another angio-*something* - *angioblastic, hemangioblastic, etc.* The *angio* keeps coming up and causes confusion. It might be best to simply call these tumors *highly vascular meningiomas.* The vessels in such tumors vary from small capillaries to large-sized vessels. They simply reflect vascularity, no more.

A question surgeons continue to ask is, "Is this a malignant meningioma?" I know this is not the place to get deeply into the issue of what constitutes malignancy. All I am saying is that we might try to address this more clearly in the Blue Book. The literature in recent years has succeeded in defining atypical meningiomas as tumors which you can expect to have an increased recurrence rate. This designation includes tumors that have areas of dense cellularity, so-called *sheeting*, foci of spontaneous necrosis, and a significant increase in mitotic activity. On the other hand, when one sees frank anaplasia, extensive necrosis, high cellularity, a very high mitotic index, etc., then application of the term *atypical* is fooling ourselves. Surely such a tumor will, at minimum, recur.

We cannot characterize a malignant meningioma merely by the presence of nearby satellite nodules because these are seen also in benign tumors. Alternatively, if there are metastases, the tumor *must be* malignant. We cannot rely on that sort of useless definition of malignancy.

Brain invasion is perhaps a useful criterion. I see it with some regularity on a microscopic scale when adequate pieces of tissue are available from the brain-tumor interface. There is, however, confusion at the microscopic level. What about perivascular infiltration - does that represent brain invasion? I think the most important prognostic index of invasion is perhaps the gross brain invasion that is apparent to the surgeon.

My final comment relates to fibrous histiocytomas. Earlier, Dr. Kepes showed elegantly that tumors of the meninges may exhibit both epithelial and mesenchymal characteristics. I am fully prepared to accept the fact that what we are calling hemangiopericytoma may be an extreme manifestation of mesenchymal differentiating capacity, and that in rare instances we may see examples at the interface between meningioma and hemangiopericytoma. But I see the fibrous histiocytoma as presenting a more difficult problem. We all recognize *type ordinaire* meningiomas that have a beautiful storiform pattern. However, when this pattern is dominant, unless we apply antisera such as EMA, we may not be able to tell whether the lesion is a fibrous meningioma or a real fibroblastic tumor. I would point out that we are considering a classification that can be used worldwide. So, unless we have a definitive way of distinguishing malignant or ordinary fibrous histiocytoma from meningioma by routine diagnostic methods, what benefit will we provide by illustrating and distinguishing this lesion?

DR. ZÜLCH: I want to come back to one point which was briefly mentioned earlier and I will introduce it by an anecdote.

At the end of a session at a meeting of general pathologists who had been shown a series of histologic slides and were asked after each what was the age, sex, and the location of the tumor, we showed an almost entirely calcareous type of psammomatous meningioma. The response was that it came from a lady between 15 and 60 years of age and that it had been located in the thoracic part of the vertebral column. This answer has a probability of about 70%. This certainly has something to do with the high preponderance of these tumors in females. Some series have reported a ratio as high as 1:20. Ours was one male to six females. There is also another common location for these psammomatous tumors, at the base of the skull near the sphenoid wing of the temporal bone. The reason for the psammomatous appearance is unknown. It might be related to vascular supply, or in the spinal region it might be due to compression.

We have in our collection about 10,000 tumors which were reclassified about eight years ago by a Japanese group of guest researchers. Among these tumors there were about 1500 meningiomas. When we classified them according to location and sex, we found some differences between those near the sagittal sinus and others in the anterior, middle, and posterior fossae which appeared to be age dependent. Moreover, with increasing age, there was a greater proportion of men.

These figures were never published, unfortunately. They should be worked up again to look for further clues.

Dr. Bruner presented some very interesting facts about the endocrine and immunochemical relationships which may exist. Nevertheless, we should not lose sight of the observations of the clinician, as a trigger to our further research. Do not forget that the first pathology laboratories were all in surgical clinics, and there was daily contact between the pathologist and the surgeon. We should continue to maintain close contact with our neurosurgeons and their important clinical obseervations.

DR. BRUNER: Dr. Kepes has been very helpful in summarizing some points for our consideration. These include the suggestion that hemangioblastic meningioma be omitted and reclassified with hemangioblastomas.

Another suggestion was that psammomatous meningioma become a subcategory of meningotheliomatous meningioma. Another point for consideration was the addition of secretory meningioma as a separate category, presumably as one of the variants. Another subgroup, meningioma with massive lymphoplasmacellular infiltrate, might be added, presumably as a subcategory of meningioma.

DR. KEPES: What about also including meningioangiomatosis as another subgroup of meningioma?

DR. BRUNER: These are definitely items for consideration which may not be fully discussed at this meeting but should be considered for the WHO Blue Book at a later date.

DR. GILLES: I would like to remind you that we are neuropathologists and the external community expects us to know brain tumors in their entirety. we are expected to know what the clinical evolution is, what the response is to various kinds of therapy and the significance of the *neuroimages*. We are expected to go to the operating room and to advise in the course of the operation.

In our classification of brain tumors, we may diminish that image somewhat if we base the classification entirely upon hypotheses about speculative relationships between tumor cells and cells in the developing nervous system. The same will be true if we divorce those classifications from concepts about etiology, knowledge about pathogenesis, or specific statements about prognosis.

I heard again, earlier at this meeting, the curious statement that we are allowed to divorce in some vague abstract way histologic malignancy from what the clinician and the patient needs to know about biologic malignancy. I think this is a mistake. I do not know how many of you have been privileged to talk to a teenager with a brain tumor and his parents. Life for that teenager, and his questions, are far simpler than these issues we have discussed over the last two days.

The first question is always about length of survival, and the second question is, "Why did it happen to me?" Hopefully, a number of experts in this room are approaching answers to the second question. I think it behooves the rest of us to concentrate upon what is going to happen, and how long it is going to take. We need to provide real answers with minimal amounts of variance about survival expectations for people with brain tumors. We need classifications that incorporate measurement, particularly measurement of survival.

What do I mean by survival? We have seen during this meeting several simple univariant survival curves, comparing different classes of tumor. Most of these univariant survival curves, even though put on the same image, are not comparable because they did not control simultaneously for the effects of location, age, sex and a large number of other clinical variables.

We saw in the paper of Dr. Daumas-Duport, from the Mayo Clinic, a series of survival curves and two other very important points. The survival curves from the Mayo Clinic were built upon Cox regression models and, in fact, those survival curves were comparable because the effects of site, age, sex and other clinical variables were simultaneously taken into consideration.

There were two other points in that same paper that I found very important. One was the attempt to reduce the tremendous number of histologic features that we try to consider in our estimates of survival. You saw the four very straightforward histologic features that they used in making their judgments. The second important thing in that paper was the measurement of the inter-observer variation for each of those features.

So, instead of damning the statisticians, as some of us tend to do, or assigning the entire responsibility for providing survival curves to the statisticians, as others would have us do, it is our responsibility to take over and understand enough of these modern statistical methods to decide whether the statisticians are providing us with accurate and adequate information.

DR. HASSOUN: I completely agree with Dr. Scheithauer regarding the term *angiomatous*. I think we have to get rid of it. The use of the phrase *highly vascular meningiomas* would be more appropriate.

My second comment concerns hormone receptors in meningiomas. I have tried very recently to demonstrate hormone receptors in a small series of meningiomas by using antibodies against estrogen receptors and progesterone receptors obtained from Abbott Laboratories and Transbio, exactly as had previously been done with breast cancers. Unfortunately, it did not work at all on frozen sections of those meningiomas contrary to what was observed with the breast cancers. It was not because of techniques but perhaps due to a difference between the epitopes of these proteins.

Thirdly, I have a comment about the predictability of malignancy of meningiomas. The only case we have observed of metastatic meningioma was in a child, who developed lymph node metastases six months after an operation for an occipital meningioma. These metastases were absolutely devoid of any mitotic activity and cell pleiomorphism. The primary brain tumor was a very classical meningioma. The metastases had occurred in spite of this benign histologic appearance.

My fourth and last comment concerns malignant fibrous histiocytoma. I am not sure that this tumor is a genuine biphasic tumor. We have observed three cases which developed in the brain parenchyma or in the meninges. They are very classical sarcomas consisting only of malignant myofibroblasts. These features correspond to what is described in other sites and I do not think that they have anything to do with true meningiomas. They are something else.

DR. FIELDS: When you think about risk versus benefit, it seems surprising to me that in all of the discussion of meningiomas during this session, no one has mentioned radiation-induced meningiomas. It is not uncommon for children who have had scalp lesions irradiated early in life to later develop such a tumor. I was surprised that nobody has mentioned this.

DR. RUBINSTEIN: I think I mentioned one case.

DR. KEPES: If I had been allowed more time, a lot of things about chromosomes, familial occurrence, and childhood meningiomas would have been covered. I think this is an important issue with respect to etiology. In a large-scale study in the late 1940's of Israeli immigrants who had received wide-field low power radiation for fungal infections of the scalp, there was an increased incidence of subsequent malignant meningiomas among them. The rate of meningiomas in that population was 4.8 times that expected in the general population. There is no question that radiation-induced meningiomas are usually bad actors.

DR. SCHEITHAUER: I have a comment to make about the issue of hormone receptors in meningiomas. We have had an opportunity recently to examine about 90 cases. We employed the immunoperoxidase method, the charcoal-binding assay method, now in use in most institutions, and a nuclear binding assay (NBA), developed by Dr. Thomas SPELSBERG of our institution. The latter is a very sensitive technique which can detect less than a hundred molecules per nucleus.

We were not surprised by one observation that mirrored what Dr. Kepes was saying, namely, that estrogen receptors are far less common than progesterone receptors. Moreover, the immunoperoxidase technique is often disappointingly negative is often disappointingly negative in the face of a positive charcoal assay or NBA. It might be asked whether these methods all measure the same thing.

Although I do not want to be nihilistic, I think the assessment of steroid receptors in meningiomas is going to be a biochemical activity. Routine immunohistology, unless we get some good antisera, is probably going to have little value in identifying receptors.

Closing Discussion

DR. SOBIN: As you can imagine, it is a rather formidable task to try to summarize this meeting, which varied greatly in its scope, from classical nosology to gene probes. I thought I would divide my summary into two parts. They are both rather short, and I will start with the first part.

When you are classifying brains
For their tumors and their names
There are problems that are many, and complex.
What's the type of neoplasia,
And its grade of cell dyscrasia
Mitotic rate, necrosis, age and sex.
Do you think they're embryonal
With a defect chromosomal,
From a hit through open window vulnerability?
Is it medulloblastoma, a pineal germinoma
Or PNET causing great hostility?
They say roses smell the same,
It's no matter what's their name,
As long as you have Blue Books and a lexicon.
I can see a great solution,
Between Lucy, Klaus, and Lucien,
We will translate all our terms to
Southeast Texican.

Well, now in a slightly less flippant vein, I would like, in summary, to consider the WHO classification in terms of its performance status. I think classifications are being tested both after as well as before they are formulated, and what we are seeing are the results of the testing of the classification after it has been published. What we should be asking now is how has the WHO classification rated? What is its state of health, what is its performance status?

I am not sure how to measure performance status. There is the Karnofsky scale. There is also another scale that can be used. It is called the Bichofsky scale. At the top, given a numerical rating of 10, you have a classification that is of worldwide use, daily used by all pathologists, clinicians, neurosurgeons, neuro-oncologists, neuroepidemiologists, and is translated into 80 languages, cited in every relevant paper, unmodified, and not criticized. It is Utopian.

At the other end of the scale, perhaps at a level of 1, there is the classification that is never modified, never cited, only ignored. Such an example is the WHO classification of lymphomas. So, we should try to rate the classification according to some scale with these characteristics. My impression is, although I am not a neuropathologist and I have been listening to you attentively during this meeting, that though the classification is not perfect, it is in very good health. It is being widely used, it is stimulating debate, and in spite of attempts at modification, it has been left intact.

There have been many suggestions on how we might improve it, but as Robin Barnard mentioned, these are merely fine tuning. There has been no suggestion of major surgery, certainly not radiotherapy, or chemotherapy, and I think those who have difficulty with it would perhaps recommend taking half an aspirin a day if they are over fifty. I do not see or hear of any major change that is being recommended.

So, after that status assessment, what would be the prescription, what should we recommend? These are only my feelings, and I would like to hear more discussion, if possible. I think, personally, we should inform WHO, the parent of the classification, and the employers of the classification, who are the users-at-large, the community of neuropathologists, neuro-oncologists, and surgeons, that the classification is in a good state of health and that it can return to work. The means for doing this will be the report of this meeting. That certainly will be the best way to inform the users-at-large.

Secondly, we should continue the ongoing testing that has been done, and that we find all of you doing. Then we must determine the impact on the classification of the newer techniques that have been brought out during this meeting. Should we either modify it or modulate it? We should be considering that question continually in all of the studies which are in progress, whether they be clinical or more in the realm of basic science.

Thirdly, we should not allow *this patient* to be lost to follow-up. We should bring it back again for a follow-up examination at a later date, and ascertain its performance status. That is the best way I can sumarize this meeting.

DR. FIELDS: I share with Dr. Sobin the sense that things should not be tampered with at the present time. We should look at certain specific issues, particularly those which have been discussed here. Then at another meeting, perhaps 18 months from now, there will be a lot of solid, new information available that may impact on our decisions regarding amendments to the WHO classification. I would like to make it clear that when I was arranging this meeting, it was my intention to use it as a springboard for further discussion and not for recommendation of a revision of the Blue Book at this time.

DR. KLEIHUES: Basically, I very much agree with Dr. Sobin's view. The Blue Book has done a tremendous job over the past ten years. It has streamlined and unified the classification and we have achieved, particularly in the German-speaking countries, an extent of consensus unimaginable ten or fifteen years ago. Nevertheless, I believe an upgrade is necessary and we should not wait too long. The first edition contains some diagnostic terms which should be excluded and there are some new tumor entities which should be added. Any unnecessary delay creates a vacuum and other people will step in.

In conclusion, a moderate upgrade which would probably affect no more than 6 or 8 tumor types should be published rapidly, within a year or two.

DR. WECHSLER: I think it is a very good idea to continue our conversations by letters, telephone, or other means of communication. This meeting was very informative but we could not discuss all the important brain tumor types, only segments of the whole field. These matters need some time. It is very important that groups should continue to meet and discuss these issues.

DR. RUBINSTEIN: I agree with almost everything that has been said here. I would tend to side with Wolfgang Wechsler rather than Paul Kleihues in advising caution before recommending a revision of the *Blue Book* which, after all, took ten years to produce, which has no references, and therefore has a life of its own. It is still valid and criticisms have been relatively minor compared with the amount of work and information that is behind it.

DR. FIELDS: I had hoped when we got together that we would come out with something at the end that would be worth circulating. Otherwise, we would not have gone to the trouble of recording and transcribing the proceedings.

DR. RORKE: As a person responsible for some of the issues at contention here, I would like to make a few comments. Those of us who are practicing pathologists dealing day-to-day with clinical problems and interacting, as Dr. Gilles indicated, not only with the families of patients, but with the clinicians who are caring for the patients, have to modify all of the information which forms the background of our medical education and our experience. We utilize the experience of those who are wiser than we are, and who have put their wisdom into textbooks, which we regard as classical guideposts, within the framework of our experience. Any classification which has been put forward has been used in this way.

Professor Zülch, in his review of the history of classifications has done a very elegant job of indicating that these have changed over a period of years. There has been no hesitation on the part of individuals to step forward and voice their dissent relative to the ideas advanced by one group or another. I do not think this is something to be feared.

We try to classify things according to the manner in which we think they are related. These may correspond to some reality in the world as we see things, or they may not. As scientists, we must try to exert as objective a point of view as feasible in finding the reality of our concepts. We use classification systems to help us order our universe, in this case, of brain tumors. If they help us, we continue to use them. If they do not, we reject them.

The proposed revision of pediatric brain tumors presented once more at this meeting has been much misunderstood. The group of us whose names are on that paper found certain difficulties when we tried to apply the WHO classification. We knew, of course, that it had been very carefully prepared by neuropathologists with many years of experience. In practice, many categories were clearly defined and useful. On the other hand, when dealing with the group of pediatric brain tumors that we are called upon to address, the classification did not fit very well. Thus, it was in that spirit that we proposed the alterations to the work of the original WHO Committee. We tried, as much as possible, to follow their proposal to avoid confusion.

Finally, a comment about the primitive neuroectodermal tumor (PNET) and the controversy which currently rages over that. I feel certain that it is not going to be settled by the people sitting in this room. However, it is important to tell you that the concepts which I outlined in my presidential address in 1982 were similar to those voiced independently by Larry Becker. He dealt with a large population of pediatric brain tumors in Canada and I had similar experience in Philadelphia. Quite independently our concepts evolved over the same period of time and we arrived at the same conclusion. In fact, he had a paper in press in *Human Pathology* at the time I was making my address to the American Association of Neuropathologists in June 1982.

Among pediatric neuropathologists, there is little or no controversy relative to this particular concept. Those of us who have had the responsibility for looking at large numbers of tumors in children have tried to explain to our clinical colleagues how the examples which are differentiated in the literature are different from each other, how they are the same, and what the biology is. We have found ourselves with a difficult problem in trying to separate all of these entities.

If the concepts that have been proposed are found to be invalid, then so be it. At this point scientific investigation is necessary to prove or disprove the validity of the proposal. If our proposal does nothing more than open the entire question of the nature of this fascinating category of tumors, and if we eventually arrive at a clear understanding of what they are, then it will have served its purpose. It does not really matter what we call them.

DR. KEPES: I have a strong feeling that some people will not like me for what I am going to say. But much as I think it is very important to come to some kind of decision on PNET, to suggest that that was the main reason to bring all of these people together, and that the first day was *the day* and the others were just added to it so that we could have an excuse to invite so many people, I think is an exaggeration. I do not think it was included in my letter of invitation from Dr. Fields that this was going to be essentially a discussion of PNET, as important as that subject may be.

Furthermore, I am in 100% agreement with Dr. Kleihues. I think that the *Blue Book*, which is now ten years old, deserves a new edition, a revised edition. The analogy that we heard, "This book is a patient and we have to see if it needs any surgery or not, and if it ain't broken, don't fix it," I think is an incorrect analogy. Why should the book be regarded as the patient? Let us look at it as a hospital that is getting a new wing or new decoration. You do not wait for a hospital to break down. We are not talking about a suffering patient; we are talking about a book.

Dr. Rubinstein is bringing out a new edition of RUSSELL and RUBINSTEIN (1989), which we are all eagerly awaiting, and I would strongly emphasize that there is absolutely nothing seriously wrong with the old edition. It is not a patient; it is just coming out in a new dress with augmented values to it, and I think this is the way we should look at the *Blue Book*.

DR. FIELDS: I hear the words *upgrade*, *new edition*, and so on, rather than "new book" and that makes abundantly good sense. There may be a number of issues discussed during these three days that bear some consideration for placement of new terms and/or new explanations in such a revision.

DR. BURGER: I wanted to put in a word on behalf of the pediatric pathologists. I am not at a pediatric hospital, so I do not qualify as such a pathologist, but I do review a substantial number of slides on a day-to-day basis for a pediatric cooperative study so I have sympathy for the pediatric pathologist confronting brain tumors in children. Although there are many entities in the WHO classification, many tumors in children do not fit into these niches. It is for this reason that Dr. Rorke assembled the group that met at Niagara-on-the-Lake. The World Health Organization classification is one largely directed at adult tumors, and the purpose of that meeting was to modify this for children.

Fortunately, or unfortunately, it got into the PNET issue which obviously generated controversy and overshadowed the need to rethink the classification of other pediatric brain tumors. It is easy to classify some neoplasms in children as astrocytomas, but these come in many varieties. You see these lesions with very distinctive patterns and realize that there are very distinct entities, which cannot always that can be accommodated by the WHO system. Yes, it may be astrocytic and GFAP positive, but it is quite unique and the term *astrocytoma* does not do it justice.

So there really, to my knowledge, has not been a systematic study of pediatric astrocytic tumors with a view to categorizing these lesions. I think this void must be filled. The purpose of that meeting at Niagara was really quite laudable and we need to expand it in the direction of the gliomas which are not really very well defined by the WHO classification.

DR. FIELDS: I find all of these comments very helpful in steering me in a direction that we ought to go. Since I am not a member of this club of neuropathologists, perhaps I am in a better position to accomplish some of the things that have been suggested. I will take the responsibility for being the focal point for the correspondence.

DR. ZÜLCH: It is much too early to prepare a new classification. The current "Blue Book" is the product of ten years of hard work. We spent a very long time trying to reach a compromise over the various discrepancies in previous classifications. I beg you to be calm and to work further on all these matters which we have been discussed here. If you come to the next meeting with the same spirit as has been evident here, that will be wonderful and we should come even closer to resolutions of many of these concerns.

DR. FIELDS: We are nearing the end of the 20th century and things are moving a lot faster than they were at the beginning of the century. I was very much taken with the comment that Paul Kleihues made earlier that if neuropathologists, as a group, do not act when they have the opportunity to do so, someone else will take the initiative and then they are not going to like it. That is an issue that needs to be addressed.

DR. GILLES: Are you going to respond to the repeated suggestions that you hold another working meeting?

DR. FIELDS: It is not likely that I will hold another meeting. The members of this group need to communicate with one another, and perhaps three members could constitute a nucleus to build on, and then they can take the responsibility for arranging another meeting. I arranged this meeting because I felt that the opportunity was at hand to do so. I want to see a product from it and I want to share that product with everyone who has been here and disseminate the information to as wide an audience as possible. What you do with it after that is your own business.

I hope that everybody leaves here with an understanding of the direction in which we want to go in the future. Let me thank all of you for responding affirmatively to my invitation to come to Houston. It has been my pleasure and the pleasure of the M. D. Anderson Cancer Center to have you here so that we could discuss these matters. I look forward to being in touch with all of you soon again.

References

Afra D, Müller W, Slowik F, Wilcke O, Budka H, Turoczy L (1983) Supratentorial lobar ependymomas: reports on the grading and survival periods in 80 cases, including 46 recurrences. Acta Neurochir 69:243-251

Ahdevaara P, Kalimo H, Törmøa T, Haltia M (1977) Differentiating intracerebral neuroblastoma: report of a case and review of the literature. Cancer 40:784-788

Al Saadi A, Latimer F (1980) Nonrandom chromosomal abnormalities in human brain tumors. Am J Hum Genet 32:61A

Al Saadi A, Latimer F (1983) Cytogenetic and clinical correlations in neoplasms of the human central nervous system. Proc Am Assoc Res 24:10

Alguacil-Garcia A, Pettigrew NM, Sima AAF (1986) Secretory meningioma: A distinct subtype of meningioma. Am J Surg Pathol 10:102-111

Allan PM, Garson JA, Harper EI, Asser U, Coakham HB, Brownell B, Kemshead JT (1983) Biological characterization and clinical applications of a monoclonal antibody recognizing an antigen restricted to neuroectodermal tissues. Int J Cancer 31:591-598

Ammirati M, Vick N, Liao Y, Ciric I, and Mikhael M (1987) Effect of the extent of surgical resection on survival and quality of life in patients with supratentorial glioblastomas and anaplastic astrocytomas. Neurosurgery 21:201-206

Auer RN, Becker LE (1983) Cerebral medulloepithelioma with bone, cartilage, and striated muscle. Light microscopic and immunohistochemical study. J Neuropathol Exp Neurol 42:256-267

Azzarelli B, Richards DE, Anton AH, Roessmann U (1977) Central neuroblastoma. Electron microscopic observations and catecholamine determinations. J Neuropathol Exp Neurol 36:384-397

Bailey OT (1985-1986) Genesis of the Percival Bailey-Cushing classification of gliomas. Pediat Neurosci 12:261-265

Bailey P (1932) Cellular types in primary tumors of the brain. In: Penfield W (ed) Cytology and cellular pathology of the nervous system, vol. 3. New York, Hoeber, pp 905-951

Bailey P (1948) Intracranial Tumors, 2nd edition. Springfield, C C Thomas

Bailey P, Cushing H (1925) Medulloblastoma cerebelli: A common type of midcerebellar glioma of childhood. Arch Neurol Psychiatr 14: 192-224

Bailey P, Cushing H (1926) A Classification of the Tumors of the Glioma Group on a Histogenetic Basis with a Correlated Study of Prognosis. Philadelphia, JB Lippincott

Banerjee AK, Blackwood W (1971) A subfrontal tumour with the features of plasmacytoma and meningioma. Acta Neuropathol (Berl) 81:84-88

Bargmann W (1943) Die Epiphysis cerebri. In: Möllendorf W (ed), Blutgefüss- und Lymphgefäss-apparat, intersekretorische Drüsen. Berlin, Springer, (Handbuch der mikroskopischen Anatomie des Menschen, Vol. 6:309-502)

Barone BM, Elvidge AR (1970) Ependymomas. A clinical survey. J Neurosurg 33:428-438

Battifora H (1975) Intracytoplasmic lumina in breast carcinoma. A helpful histopathologic feature. Arch Pathol 99:614-617

Becker LE, Hinton D (1983) Primitive neuroectodermal tumors of the central nervous system. Human Pathol 14:538-550

Becker LE, Hinton D (1986) Primitive neuroepithelial tumors of the central nervous sytem. In: Finegold M (ed) Pathology of Neoplasia in Children and Adolescents. Philadelphia, WB Saunders, pp.397-418

Becker LE, Halliday WC (1987) Central nervous system tumors of childhood. Perspect Pediatr Pathol 10:86-134

Bennett JP Jr, Rubinstein LJ (1984) The biological behavior of primary cerebral neuroblastoma: a reappraisal of the clinical course in a series of 70 cases. Ann Neurol 16:21-27

Berard M, Tripier MF, Choux R, Chrestian MA, Hassoun J, Toga M (1978) Étude ultrastructurale des corps hyalins d'un meningiome pseudoepithelial. Acta Neuropathol (Berl) 42:59-62

Berger MS, Edwards MSB, Wara WM, Levin VA, Wilson CB (1983) Primary cerebral neuroblastoma: long-term follow-up review and therapeutic guidelines. J Neurosurg 59:418-423

Best P (1973) A medulloblastoma-like tumor with melanin formation. J Pathol Bacteriol 110: 109-111

Biggs PJ, Powers JM (1984) Neuroblastic medulloblastoma with abundant cytoplasmic actin filaments. Arch Pathol Lab Med 108:326-329

Bigner SH, Mark J, Mahaley MS, Jr, Bigner DD (1984) Patterns of the early, gross chromosomal changes in malignant human gliomas. Hereditas 101:103-113

Bigner SH, Mark J, Bullard DE, Mahaley MS, Jr, Bigner DD (1986) Chromosomal evolution in malignant human gliomas starts with specific and usually numerical deviations. Cancer Genet Cytogenet 22:121-135

Bigner SH, Mark J, Friedman HS, Biegel JA, Bigner DD (1988a) Structural chromosomal abnormalities in human medulloblastoma. Cancer Genet Cytogenet 30:91-101

Bigner SH, Mark J, Burger PC, Mahaley MS, Bullard DE, Muhlbaier LH, Bigner DD (1988b) Specific chromosomal abnormalities in malignant human gliomas. Cancer Res 48:405-411

Bigner SH, Vogelstein B, Bigner DD (1988c) Chromosomal abnormalities and gene amplification in malignant gliomas. ISI Atlas of Science: Biochemistry (in press)

Bigner SH, Burger PC, Wong AJ, Werner MH, Hamilton SR, Muhlbaier LW, Vogelstein B, Bigner DD (1988d) Gene amplification in malignant human gliomas: Clinical and histopathologic aspects. J Neuropath Exp Neurol 47:191-205

Bigner SH, Vogelstein B, Burger PC, Schold SC, Bigner DD (1988e) Amplification of the epidermal growth factor receptor gene is related to histologic grade of human gliomas (submitted)

Bingas B (1964) Das monstrozelluläre Sarkom des Gihirns. Arch Psychiatr Nervenkr 205:223-236

Blankenstein MA, Blaauw G, Lamberts SWJ (1984) Progestin and estrogen receptors in human meningioma. Clin Neuropharmacol 7:363-367

Boesel CP, Suhan JP, Bradel EJ (1978) Ultrastructure of primitive neuroectodermal neoplasms of the central nervous system. Cancer 42:194-201

Bonnin JM, Rubinstein LJ (1984) Immunohistochemistry of central nervous system tumors. Its contributions to neurosurgical diagnosis. Neurosurg 60:1121-1133

Bonnin JM, Perentes E (1988) Retinal S-antigen immunoreactivity in medulloblastomas. Acta Neuropathol (Berl) 76:204-207

Bookwalter JW, Selker RG, Schiffer L, Randall M, Lannuzzi D, Kristofik M (1986) Brain-tumor cell kinetics correlated with survival. J Neurosurg 65:795-798

Bourdon MA, Wikstrand CJ, Furthmayr H, Matthews TJ, Bigner DD (1983) Human glioma-mesenchymal extracellular matrix antigen defined by monoclonal antibody. Cancer Res 43: 2796-2805

Bourne SP, Coakham HB, Kemshead JT, Brownell B, Allan PM, Davies AG (1986) Chap 7. The role of monoclonal antibodies in the diagnosis and characterization of medulloblastoma. In: Zeltzer PM, Pochedly C (eds) Medulloblastoma in Children: New Concepts in Tumor Biology, Diagnosis and Treatment. New York, Praeger

Bright R (1831) Reports of medical cases, selected with a view of illustrating the symptoms and cure of disease by a reference to morbid anatomy. London, Longman 2:342-347

Broders AC (1926) Carcinoma: grading and practical application. Arch Path Lab Med 2:376-381

Brodeur GM, Seeger RC, Schwab M, Varmus HE, Bishop JM (1984) Amplification of N-myc in untreated human neuroblastomas correlates with advanced disease stage. Science 224: 1121-1124

Broggi C, Franzini A, Ferraresi S, Sberna M, Servello D, Cajola L, Allegranza A (1988) Cell kinetics and multi-modal prognostic evaluation in glial tumors investigated by serial stereotactic biopsy. Acta Neurochir 94:53-56

Brucher JM (1987) Classification of tumours of the nervous system. In: Grundmann E (ed): Cancer Campaign. Vol.10. Experimental Neurooncology, Brain Tumor and Pain Therapy. Stuttgart, Gustav Fischer pp.63-67

Brucher JM, Dalesio O, Solbu G (1987) Prospective analysis of grade III and IV gliomas, with consideration of histologic classification. (An EORTC Brain Tumor Group Study) In: Chatel M, Darecel F, Pecker J (eds) Brain Oncology. Dordrecht, Martinus Nijhoff, pp. 237-242

Brun A (1965) The subpial granular layer of the foetal cerebral cortex in man. Its ontogeny and significance in congenital cortical malformations. Acta Pathol Microbiol Scand Suppl 179:1-98

Budka H (1980) Hyaline inclusions (pseudopsammoma bodies) in meningiomas: immunocytochemical demonstration of epithelial-like secretion of secretory component and immunoglobulins A and M. Acta Neuropathol (Berl) 39:575-583

Burger PC, Vogel FS (1976) Surgical Pathology of the Nervous System and its Coverings, 1st Ed. New York, John Wiley and Sons, p 93

Burger PC, Vollmer RT (1985) Histologic factors of prognostic significance in the glioblastoma multiforme. Cancer 56: 1179-1186

Burger PC, Green SB (1987) Patient age, histologic features, and length of survival in patients with glioblastoma multiforme. Cancer 59:1617-162

Burger PC, Vogel FS, Green SB, Strike TA (1985) Glioblastoma multiforme and anaplastic astrocytoma. Pathologic criteria and prognostic implications. Cancer 56: 1106-1111

Burger PC, Shibata T, Kleihues P (1986) The use of the monoclonal antibody Ki-67 in the identification of proliferating cells: Application to surgical neuropathology. Am J Surg Pathol 10:611-617

Burger PC Rawlings CE Cox EB, McLendon RE, Schold SC, Bullard DE (1987a) Clinicopathologic correlation in the oligodendroglioma. Cancer 59:1345-1352

Burger PC, Grahmann FC, Bliestle A, Kleihues P (1987b) Differentiation in the medulloblastoma. A histological and immunohistochemical study. Acta Neuropathol (Berl) 73:115-123

Burger PC, Shibata T, Kleihues P (1988) Proliferation markers for neoplasms of the nervous system. In: DeLellis RA (ed) Advances in Immunohistochemistry, Raven Press, New York, pp 301-316

Camins MB, Cravioto HM, Epstein F, Ransohoff J (1980) Medulloblastma: An ultrastructural study - Evidence for astrocytic and neuronal differentiation. Neurosurgery 6:398-411

Campbell AN, Chan HSL, Becker LE, Daneman A, Park TS, Hoffman HJ (1984) Extracranial metastases in childhood primary intracranial tumors. A report of 21 cases and review of the literature. Cancer 53:974-981

Carden MJ, Schlaepfer WW, Lee VM-Y (1985) The structure, biochemical properties, and immunogenicity of neurofilament peripheral regions are determined by phosphorylation state. J Biol Chem 260:9805-9817

Carrel S, Accolla RS, Carmagnola AL, Mach J-P (1980) Common human melanoma-associated antigen(s) detected by monoclonal antibodies. Cancer Res 40:2523-2528

Challa VR, Moody DM, Marshall RB, Kelley DL Jr (1980) The vascular component in meningiomas associated with severe cerebral edema. Neurosurgery 7:363-368

Chen WYK, Kepes JJ, Teglbjaerg PS (1985) Intracellular mucoid changes in tumor cells of meningiomas: a manifestation of polyvinylpyrrolidone (PVP) effect on tissues with mesenchymal characteristics. J Neuropathol Exp Neurol 44:606-616

Cheng L, Binder SW, Fu Y-S, Lewin KJ (1988) Methods in laboratory investigation. Demonstration of estrogen receptors by monoclonal antibody in formalin-fixed breast tumors. Lab Invest 58:346-352

Cho KG, Hoshino T, Nagashima T, Murovic JA, Wilson CB (1986) The prediction of tumor doubling time in recurrent meningiomas: Cell kinetics study with bromodeoxyuridine labeling. J Neurosurg 65:790-794

Cho KG, Hoshino T, Nagashima T, Davis RL (1987) Proliferative potential of malignant and non-malignant meningiomas. In: Chatel M, Darcel F, Pecker J (eds) Brain Oncology. Dortrecht, Martinus Nijhoff, pp.79-81

Chow CW, Klug GL, Lewis EA (1988) Subependymal giant cell astrocytoma in children: an unusual discrepancy between histology and clinical features. J Neurosurg 68:890-898

Christensen D, Laursen H, Klinken L (1983) Prediction of recurrence in meningiomas after surgical treatment. Acta Neuropath (Berl) 611:130-134

Clara M (1959) Das Nervensystem des Menschen, 3rd edition. Leipzig, Barth

Clarke BF, Parsons H (1951) An embryological tumor of retinal anlage involving the skull. Cancer 4:78-85

Cleland J (1864) Description of two tumours adherent to the deep surface of the dura mater. Glasg Med J 11:158-159

Coakham HB, Garson JA, Brownell B, Allan PM, Harper EI, Kemshead JT (1984) The use of monoclonal antibodies in the immunohistological diagnosis of cerebral and spinal tumors. S Afr J Surg 22:13-22

Coakham HB, Garson JA, Allan PM (1985) Immunohistological diagnosis of central nervous system tumors using a monoclonal antibody panel. J Clin Patho 38:165-173

Coffin CM, Mukai K, Dehner LP (1983) Glial differentiation in medulloblastomas. Histogenetic insight, glial reaction, or invasion of brain? Am J Surg Pathol 7:555-565

Cohadon F, Aouad N, Rougier A, Vital C, Rivel J, Dartigues JF (1985) Histologic and non-histologic factors correlated with survival time in supratentorial astrocytic tumors. J Neuro-Oncol 3:105-111

Courriere P, Tremoulet M, Eche N, Armand J-P (1985) Hormonal steroid receptors in intracranial tumours and their relevance in hormone therapy. Eur J Cancer Clin Oncol 21:711-714

Cox LB (1933) The cytology of the glioma group with special reference to the inclusion of cells derived from invaded tissue. Am J Path 939-998

Critchley M (1986) The Citadel of the Senses. New York, Raven Press p. 176

Crompton MR,Gautier-Smith PC (1970) The prediction of recurrence in meningiomas. J. Neurol Neurosurg Psychiatry 33: 80-87

Cruz-Sanchez FF, Haustein J, Rossi ML, Cervos-Navarro J, Hughes JT (1988) Ependymoblastoma: a histological, immunohistological and ultrastructural study of five cases. Histopathol 12:17-27

Cushing H (1922) The meningiomas (dural endotheliomas). Their source and favoured seats of origin. Brain 45:282-316

Cushing H, Eisenhardt L (1938) Meningiomas, Their Classification, Regional Behaviour, Life History, and Surgical End Results. Springfield, Thomas, pp 43-44

Cushing H (1930) Experiences with the cerebellar medulloblastoma. A critical review. Acta Pathol Microbiol Scand 7:1-86

Danova M, Riccardi A, Gaetani F, Wilson GD, Mazzini G, Brugnatelli S, Buttini R, Butti G, Ucci G, Paoletti P, Ascari E (1988) Cell kinetics of human brain tumors: In vitro study with bromodeoxyuridine and flow cytometry. Eur J Cancer Clin Oncol 24:873-880

Dastur DK (1982) Cerebral ganglioglio-neuroblastoma: an unusual brain tumour of the neuron series. J Neurol Neurosurg Psychiatr 45:139-142

Daumas-Duport C, Szikla G (1981) Délimitation et configuration spatiale des gliomes cérébraux. Données histologiques, incidences therapeutiques. Neurochir 27:273-284

Daumas-Duport C, Monsaigeon V, Szikla G (1982) Serial stereotactic biopsies: a double histologic code of gliomas according to malignancy and 3D configuration as an aide to therapeutic decision and assessment of results. Appl Neurophysiol 45:431-437

Daumas-Duport C, Scheithauer BW, Kelly PG (1987) A histologic and cytologic method for the spatial definition of gliomas. Mayo Clin Proc 62:435-449

Daumas-Duport C, Scheithauer B, O'Fallon J, Kelly P (1988) Grading of astrocytomas: a simple and reproducible grading method. Cancer 62:2152-2165

De la Motte S, Flickinger J, Longgood RM (1986) Histopathologic features predicting recurrence of meningiomas following subtotal resection. Am J Surg Pathol 10:836-843

de Chadarévian J-P, Montes JL, O'Gorman AM, Freeman CR (1987) Maturation of cerebellar neuroblastoma into ganglioneuroma with melanosis. A histologic, immunocytochemical, and ultrastructural study. Cancer 59:69-76

Deck JHN (1969) Cerebral medulloepithelioma with maturation into ependymal cells and ganglion cells. J Neuropathol Exp Neurol 28:442-454

Deckert M, Reifenberger G, Wechsler W (1989) Determination of the proliferative potential of human brain tumors by the monoclonal antibody Ki-67. J Cancer Res Clin Oncol (in press)

Dehner LP (1986) Peripheral and central primitive neuroectodermal tumors: A nosologic concept seeking a consensus. Arch Pathol Lab Med 110:997-1005

Dehner LP (1987) Primitive neuroectodermal tumors (letter). Arch Pathol Lab Med 111:311-312

Dehner LP, Sibley RK, Sauk JJ et al (1979) Malignant melanotic neuroectodermal tumor of infancy: A clinical, pathologic, ultrastructural and tissue culture study. Cancer 43:1389-1410

del Rio Hortega P (1930) Para el mejor conocimiento histologico de los meningoexotheliomas. Arch Esp Oncol 1:477-570

del Rio Hortega P (1932) Estructura y sistematisación de los gliomas y paragliomas. Arch Esp Oncol 2:411-678

Derynck R, Goeddel DV, Ullrich A, Gutterman JU, Williams RD, Bringman PS, Berger WH (1987) Synthesis of messenger RNAs for transforming growth factors α and β and the epidermal growth factor by human tumors. Cancer Res 47:707-712

Dohrmann GJ, Farwell JR, Flannery JT (1976) Ependymomas and ependymoblastomas in children. J Neurosurg 45:273-283

Donner L, Triche TJ, Israel MA et al (1985) A panel of monoclonal antibodies which discriminate neuroblastoma from Ewing's sarcoma, rhabdomyosarcoma, neuroepithelioma and hematopoietic malignancies. In: Evans A (ed) Advances in Neuroblastoma Research. New York, Alan R Liss pp.347-366

Drut R, Jones MC (1983) Melanotic medulloblastoma of infancy: A case report with immunohistochemical study and literature review. 7:53-62

Eng LF, DeArmond SJ (1982) Immunocytochemical studies of astrocytes in normal development and disease. In: Fedoroff S, Hertz L (eds), Advances in Cellular Neurobiology. New York, Academic Press, pp 143-165

Eng LF, DeArmond SJ (1983) Immunochemistry of the glial fibrillary acidic (GFA) protein. In: Zimmerman HM (ed) Progress in Neuropathology, Vol. V. New York, Raven Press, pp 19-39

Eng LF, Lee YL, Miles LEM (1976) Measurement of glial fibrillary acidic protein by a two-site immunoradiometric assay. Anal Biochem 71:243-259

Ermel AE (1974) Histogenesis of angiomatous areas in meningiomas. An electron microscopic study. Path Europ 9:217-231

Fasekas JT (1977) Treatment of grade 1 and 2 astrocytomas. The role of radiotherapy. Int J Radiation Oncol Biol Phy 2:661-666

Filmus J, Follak MN, Cairncross JG, Buick RN (1985) Amplified, overexpressed and rearranged epidermal growth factor receptor gene in a human astrocytoma cell line. Biochem Biophys Res Comm 131: 207-215

Foerster O, Gagel O (1931) Ein Fall von sogenanntem Gliom des Nervus opticus-Spongioblastoma multiforme ganglioides. Z Gesamte Neurol Psychiatr 136:335-366

Font RL, Croxatto JO (1980) Intracellular inclusions in meningothelial meningioma: A histochemical and ultrastructural study. J Neuropathol Exp Neurol 39:575-583

Fowler M (1968) Embryonic ependymoma arising in a cerebral hemisphere. Cancer 21:1150-1156

Frankel AE, Ring DB, Tringale F, Hsieh-Ma ST (1985) Tissue distribution of breast cancer-associated antigens defined by monoclonal antibodies. J Biol Response Mod 4:273-286

Fredman P, Magnani JL, Nirenberg M, Ginsburg V (1984) Monoclonal antibody A2B5 reacts with many gangliosides in neuronal tissue. Arch Biochem Biophys 233:661-666

Friedman HS, Burger PC, Bigner SH, Trojanowski JQ, Wikstrand CJ, Halperin EC, Bigner DD (1985) Establishment and characterization of the human medulloblastoma cell line and transplantable xenograft D283 med. J Neuropath Exp Neurol 44:592-605

Friedman HS, Burger PC, Bigner SH, Trojanowski JQ, Brodeur GM, He X, Wikstrand CJ, Kurtzberg J, Berens ME, Halperin EC, Bigner DD (1988) Phenotypic and genotypic analysis of a human medulloblastoma cell line and transplantable xenograft (D341 med) demonstrating amplification of c-myc. Amer J of Pathol, 139:472-484

Friedman NB (1947) Germinoma of the pineal. Its identity with germinoma (seminoma of the testis). Cancer Res 7:363-368

Friend SH, Dryja TP, Weinberg RA (1988) Oncogenes and tumor suppressing genes. New Eng J Med 318:618-622

Fujita S (1963) The matrix cell and cytogenesis in the developing central nervous system. J Comp Neurol 120:37-42

Fujita S (1986) Transitory differentiation of matrix cells and its functional role in the morphogenesis of the developing vertebrate CNS. In: Okada TS, Kodoh H (eds) Commitment and Instability in Cell Differentiation, Vol 20. New York, Academic Press, pp 223-242

Fukui M, Iwaki T, Sawa H, Inoue T, Takeshita I, Kitamura K (1986) Proliferative activity of meningiomas as evaluated by bromodeoxyuridine uptake examination. Acta Neurochir (Wien) 81:135-141

Fulling KH, Garcia DM (1985) Anaplastic astrocytoma of the adult cerebrum. Prognostic value of histologic features. Cancer 55:928-931

Fülop Z, Lakos I, Basco E, Hajos F (1979) Identification of early glial elements as the precursors of Bergmann-glia: A Golgi-analysis of the developing rat cerebellar cortex. Acta Morphol Acad Sci (Hungary) 27:273-280

Gaffney CC, Sloane JP, Bradley NJ, Bloom HJG (1985) Primitive neuroectodermal tumors of the cerebrum. Pathology and treatment. J Neuro-Oncol 3:23-33

Gambarelli D, Hassoun J, Choux M, Toga M (1982) Complex cerebral tumor with evidence of neuronal, glial, and Schwann cell differentiation: a histologic, immunocytochemical and ultrastructural study. Cancer 49:1420-1428

Garcia DM, Fulling KH, Marks JE (1985) The value of radiation therapy in addition to surgery for astrocytomas of the adult cerebrum. Cancer 55:919-927

Gehan EA, Walker MD (1977) Prognostic factors for patients with brain tumors. Natl Cancer Inst Monogr 46:189-195

Gerdes J, Schwab U, Lemke H, Stein H (1983) Production of a mouse monoclonal antibody reactive with a human nuclear antigen associated with cell proliferation. Int J Cancer 31:13-20

Gerdes J, Lemke H, Baisch H, Wacker HH, Schwab U, Stein H (1984) Cell cycle analysis of a cell proliferation-associated human nuclear antigen defined by the monclonal antibody Ki-67. J Immunol 133: 1710-1715

Ghandour MS, Labourdette G, Vincendon G, Gombos G (1981) A biochemical and immunohistological study of S100 protein in developing rat cerebellum. Dev Neurosci 4:98-109

Ghandour MS, Langley OK, Clos L (1983) Immunohistochemical and biochemical approaches to the development of neuroglia in the CNS, with special reference to cerebellum. Int J Devel Neurosci 1:411-425

Giangaspero F, Doglioni C, Rivano MT, Fileri S, Gerdes J, Stein H (1987) Growth fraction in human brain tumors defined by the monclonal antibody Ki-67. Acta Neuropathol (Berl) 74:179-182

Gilles FH, Leviton A, Hedley-White ET, Jasnow M (1983) Childhood brain tumor update. Human Pathol 15:834-845

Giordana MT, Mauro A, Migheli A, Schiffer D (1983) Contributions of immunohistochemistry to the problem of differentiation in medulloblastoma. Ital J Neurol Sci 4:411-415

Glasauer FE, Yuan RHP (1963) Intracranial tumors with extracranial metastases: Case report and review of the literature. J Neurosurg 20:474-493

Globus JH, Strauss I (1925) Spongioblastoma multiforme. Arch Neurol Psych 14:139-151

Globus JH, Silbert S (1931) Pinealomas. Arch Neurol 25:937-985

Gottschalk J, Szymas J (1987) Bedeutung der immunhistochemie für die Neuroonkologie. VI Mitteilung: Vorkommen, Lokalisation und Verteilungsmuster des sauren Gliafaserproteins (GFAP) in 820 Tumoren. Zentralbl allg Pathol pathol Anat 133:319-330

Gould VE (1986) Histogenesis and differentiation: A reevaluation of these concepts as criteria for the classification of tumors. Human Pathol 17:212-215

Gratzner HG (1982) Monoclonal antibody to 5-bromo- and 5-iododesoxyuridine: A new reagent for detection of DNA replication. Science 218:474-476

Griffin CA, Hawkins AL, Packer RJ, Rorke LB, Emanuel BS (1988) Chromosome abnormalities in pediatric brain tumors. Cancer Res 48:178-180

Grunert V (1973) Astrocytomas of the cerebrum. In: Modern Aspects of Neurosurgery, Vol 3. Internat Congr Ser No. 287. New York , American Elsevier, p. 108

Halliday WC, Yeger H, Duwe GF, Phillips MJ (1985) Intermediate filaments in meningiomas. J Neuropathol Exp Neurol 44:617-623

Hanjan SNS, Kearney JF, Cooper MD (1982) A monoclonal antibody (MMA) that identifies a differentiation antigen on human myelomoncytic cells. Clin Immunol Immunopathol 23:172-188

Hart MN, Earle KM (1973) Primitive neuroectodermal tumors of the brain in children. Cancer 32:890-897

Hassoun J, Gambarelli D, Grisoli F, Pellet W, Salamon G, Pellissier JF, Toga M (1982) Central neurocytoma. An electron microscopic study of two cases. Acta Neuropathol (Berl) 56:151-156

Hassoun J, Gambarelli D, Peragut JC, Toga M (1983) Specific ultrastructural markers of human pinealomas. A study of four cases. Acta Neuropathol (Berl) 62:31-40

Hayward E, Whitwell H, Paul KS, Barnes DM (1984) Steroid receptors in human meningioma. Clin Neurophamacol 7:351-356

He X, Skapek SX, Wikstrand CJ, Friedman HS, Trojanowski JQ, Kemshead JT, Coakham HB, Bigner SH, Bigner DD (1989) Phenotypic analysis of four human medulloblastoma cell lines and transplantable xenografts. J Neuropath Exp Neurol 48:#1

Herman MM, Sipe JC, Rubinstein LJ, VandenBerg SR, Spence AM, Vraa-Jensen J. (1975) An experimental mouse testicular teratoma as a model for neuroepithelial neoplasia and differentiation. II. Electron microscopy. Am J Pathol 81:421-444

Herman MM, Rubinstein LJ (1984) Divergent glial and neuronal differentiation in a cerebellar medulloblastoma in an organ culture system: in vitro occurrence of synaptic ribbons. Acta Neuropathol (Berl) 65:10-24

Herpers MJHM, Budka H (1984) Glial fibrillary acidic protein (GFAP) in oligodendroglial tumors: Gliofibrillary oligodendroglioma and transitional oligoastrocytoma as subtypes of oligodendroglioma. Acta Neuropathol (Berl) 64:265-272

Herpers MJHM, Budka H (1985) Primitive neuroectodermal tumors including the medulloblastoma: glial differentiation signaled by immunoreactivity for GFAP is restricted to the pure desmoplastic medulloblastoma ("arachnoidal sarcoma of the cerebellum"). Clin Neuropathol 4:12-18

Herrick MK, Rubinstein LJ (1979) The cytological differentiating potential of pineal parenchymal neoplasms (true pinealomas). Brain 102:289-320

Hewing M (1980) Synaptic ribbons in the pineal system of normal and light deprived golden hamsters. Anat Embryol 159:71-80

Hinton D, Mobbs EG, Sima AA, Hanna W (1983) Steroid receptors in meningiomas: A histochemical and biochemical study. Acta Neuropathol (Berl) 62:134-140

Hirano A (1971) Electron microscopy in neuropathology. In: Zimmerman HM (ed) Progress in Neuropathology. London, Heineman Medical Books Ltd, pp 1-61

Hirano A, Shin WY (1979) Unattached presynaptic terminals in a cerebellar neuroblastoma in the human. Neuropathol Appl Neurobiol 5:63-70

His W (1904) Die Entwicklung des menschlichen Gehirns. Leipzig, Herzel

Hitchcock E, Morris CS (1987) Immunocytochemistry of intracranial meningiomas. J Neuro-oncol 5:357-368

Holden J, Dolman CL, Churg A (1987) Immunohistochemistry of meningiomas including the angioblastic type. J Neuropathol Exp Neurol 46:50-56

Horten BC, Rubinstein LJ (1976) Primary cerebral neuroblastomas. A clinicopathological study of 35 cases. Brain 99:735-756

Horten BD, Urich H, Rubinstein LJ, Montague SR (1977) The angioblastic meningioma: A reappraisal of a morphological problem. J. Neurol Sci, 31:387-410

Horten BC, Urich H, Stefoski D (1979) Meningiomas with conspicuous plasma cell-lymphocytic components. A report of five cases. Cancer 43:258-264

Hoshino T (1987) Proliferative potential of pediatric brain tumors. Prog Exp Tumor Res 30: 31-43

Hoshino T, Sano K (1969) Radiosensitization of malignant brain tumors with bromouridine (thymidine analogue). Acta Radiol Ther Phys Biol 8:15-26

Hoshino T, Wilson CB (1979) Cell kinetic analyses of human malignant brain tumors (gliomas). Cancer 44:956-962

Hoshino T, Nagashima T, Murovic J, Levin EM, Levin VA, Rupp SM (1985) Cell kinetic studies of in situ human brain tumors with bromodeoxyuridine. Cytometry 6:627-632

Hoshino T, Nagashima T, Cho KG, Murovic JA, Hodes JE, Wilson CB, Edwards MSB, Pitts LE (1986) S-phase fraction of human brain tumors in situ measured by uptake of bromodeoxyuridine. Int J Cancer 38:369-374

Hoshino T, Nagashima T, Murovic J, Wilson CB, Edwards MSB, Gutin PH, Davis RL, DeArmond SJ (1986a) In situ cell kinetics studies on human neuroectodermal tumors with bromodeoxyuridine labeling. J Neurosurg 64:453-459

Hoshino T, Nagashima T, Murovic JA, Wilson CB, Davis RL (1986b) Proliferative potential of human meningiomas of the brain. A cell kinetics study with bromodeoxyuridine. Cancer 58:1466-1472

Humphrey PA, Wong AJ, Friedman HS, Werner MH, Vogelstein B, Bigner DD, Bigner SH (1987) Amplification and expression of the epidermal growth factor receptor (EGFR) gene in human glioma (HGL) xenografts. Proc Ann Meet Am Assoc Cancer Res 28:23 (abstract)

Humphrey PA, Wong AJ, Vogelstein B, Friedman HS, Werner MH, Bigner DD, Bigner SH (1988) Amplification and expression of the epidermal growth factor receptor gene in human glioma xenografts. Cancer Res 48:2231-2238

Ito M, Nagashima T, Hodes JE, Hoshino T (1988) The proliferative activity of human brain tumors: comparison of the labeling indices obtained with Ki-67 and with monoclonal antibody against bromodeoxyuridine. Acta Neuropathol (Berl) (in press)

Iwaki T, Takeshita I, Kufui M, Kitamura K (1987) Cell kinetics of the malignant evolution of meningothelial meningioma. Acta Neuropathol (Berl) 74:243-247

Jääkelainen HJ, Haltia M, Laassonen E, Wohlström T, Valtonen S (1985) The growth rate of intracranial meningiomas and its relation to histology. An analysis of 43 cases. Surg Neurol 24:165

Jacobsen PF, Jenkyn DJ, Papadimitriou JM (1985) Establishment of a human medulloblastoma cell line and its heterotransplantation into nude mice. J Neuropathol Exp Neurol 44:472-485

Jakob A (1928) Das Kleinhirn. In: von Möllendorff W (ed) Handbuch der mikroskopischer Anatomie des Menschen, vol. 4. Berlin, Springer

James CD, Carlbom E, Dumanski JP, Hansen M, Nordenskjold M, Collins VP, Cavenee WK (1988a) Clonal genomic alterations in glioma malignancy stages. Cancer Res 48:5546-5551

James CD, Carlbom E, Nordenskjold M, Collins VP, Cavenee WK (1988b) Mitotic recombination mapping of a locus on chromosome 17 involved in the evolution of astrocytomas (submitted)

Jay JR, McLaughlin DT, Riley KR, Martuza RL (1985) Modulation and meningioma cell growth by sex steroids in vitro. J Neurosurg 62:757-762

Jellinger K, Slowik F (1975) Histologic subtypes and prognostic problems in meningiomas. J Neurol 208:279-298

Jellinger K (1988) Histologic features predicting recurrence of meningiomas after (sub)total resection. Clin Neurosurg 7:174

Jiang Q, Lim T, Blodi FC (1984) Dual properties of cultured retinoblastoma cells. Immunohistochemical characterization of neuronal and glial markers. Exp Eye Res 39:207-215

Kalm H, Magun R (1950) Beitrag zur Klinik und pathologischen Anatomie der Pinealome. Dtsch Z Nervenheilkd 164:453-468

Kaplan MS (1982) Proliferation of subependymal cells in the adult primate CNS: differential uptake of DNA labelled precursors. J Hirnforsch 23:23-33

Karch SB, Urich H (1972) Medulloepithelioma: Definition of an entity. J Neuropathol Exp Neurol 31:27-53

Karasek M, Vollrath L (1982) "Synaptic" ribbons and spherules of the rat pineal gland: day/night changes in vitro? Exp Brain Res 46:205-208

Kawamoto T, Sato JD, Lee A, Polikoff J, Sato GH, Mendelsohn J (1983) Growth stimulation of A431 cells by epidermal growth factor: identification of

high-affinity receptors for epidermal growth factor by an anti-receptor monoclonal antibody. Proc Natl Acad Sci USA 80:1337-1341
Kemshead JT, Fritschy J, Garson JA, Allan P, Coakham H, Brown S, Asser U (1983) Monoclonal antibody UJ 127:11 detects a 220,000-240,000 kdal. glycoprotein present in a subset of neuroectodermally derived cells. Int J Cancer 1:187-195
Kepes JJ (1961) Observations on the formation of psammoma bodies and pseudo-psammoma bodies in meningiomas. J Neuropathol Exp Neurol 20:255-262
Kepes JJ (1975) The fine structure of hyaline inclusions (pseudopsammoma bodies) in meningiomas. J Neuropathol Exp Neurol 34:282-289
Kepes JJ (1982) Meningiomas. Biology, Pathology and Differential Diagnosis. Masson Monographs in Diagnostic Pathology, No. 4. New York, Masson Publ., pp. 86, 190-200
Kepes JJ, Goldware S, Leoni R (1983) Meningioma with pseudoglandular pattern. A case report. J Neuropathol Exp Neurol 42:61-68
Kepes JJ, Dunlap MD, O'Boynick P, Terreros D (1986) Meningioma with multiple rheumatoid nodules. A case report. Acta Neuropathol (Berl) 70:314-319
Kepes JJ, Chen WYK, Connors MH Vogel FS (1988) "Chordoid" meningeal tumors in young individuals with peritumoral lymphoplasmacellular infiltrates causing systemic manifestations of the Castleman syndrome. Cancer 42:391-406
Kernohan JW, Sayre GP (1952) Atlas of tumor pathology, Section X, Fascicle 35, Tumors of the central nervous system. Armed Forces Institute of Pathology, Washington
Kernohan JW, Uihlein A (1962) Sarcomas of the brain. Springfield, C C Thomas
Kernohan JW, Mabon RF, Svien HJ, Adson AW (1949) A simplified classification of gliomas. Proc Staff Meet Mayo Clin 24: 71-75
Kershman J (1938) The medulloblast and the medulloblastoma. A study of human embryos. Arch Neurol Psychiatr 40:937-967
King T, Dougherty WJ (1982) Effect of denervation on "synaptic" ribbon population in the rat pineal gland. J Neurocytol 11:19-28
Kinzler KW, Bigner SH, Bigner DD, Trent JM, Law ML, O'Brien SJ, Wong AJ, Vogelstein B (1987) Identification of amplified, highly expressed gene in a human glioma. Science 236:70-73
Kinzler KW, Ruppert JM, Bigner SH, Vogelstein B (1988) The GLI gene is a member of the Kruppel family of zinc finger proteins. Nature 332:371-374
Kleihues P, Volk B, Anagnostopoulos J, Kiessling M (1984) Morphologic evaluation of stereotactic brain tumor biopsies. Acta Neurochir (suppl) (Wien) 33:171-181
Kleihues P, Kiessling M, Janzer RC (1987) Morphological markers in neuro-oncology. In: Seifert G (ed) Current Topics in Pathology Vol.77: Morphological Tumor Markers, Springer-Verlag, Berlin Heidelberg, pp 307-338
Kleihues P, Sarioglu A, Shibata T, Yasargil MG (1989) Lack of correlation between growth fraction and patient survival in the glioblastoma multiforme (in press)
Kline KT, Damjanov I, Moriber-Katz S, Schmidek H (1979) Pineoblastoma: an electron microscopic study. Cancer 44: 1692-1699
Kock KF, Teglbjaerg, PS (1981) Meningiomas with a non-meningotheliomatous component. A new type of tumor? Acta Neuropathol (Berl) 55:199-203
Koppel H, Lewis PD (1978) Neuronal proliferation in the rat cerebellar external granular layer (Abstract). J Neuropath Exp Neurol 37:644
Korf HW, Klein DC, Zigler JS, Gery I, Schachenmayr W (1986) S-antigen-like immunoreactivity in a human pineocytoma. Acta Neuropathol (Berl) 69:165-167
Kosnik EJ, Boesel CP, Bay J, Sayers MP (1978) Primitive neuroectodermal tumors of the central nervous system in children. J Neurosurg 48:741-746

Krabbe KH (1923) The pineal gland, especially in relation to the problem of its supposed significance in sexual development. Endocrinology 7:374-414

Kramer, S. (1976) The case for subclassification of the malignant gliomas. Int J Radiation Oncology Biol Phys 1:813

Kricheff II, Becker M, Schneck SA, Taveras JM (1964a) Intracranial ependymomas: factors influencing prognosis. J Neurosurg 21:7-14

Kricheff II, Becker M, Schneck SA, Taveras JM (1964b) Intracranial ependymomas: a study of survival in 65 cases treated by surgery and irradiation. Am J Roentgenol Radium Ther Nucl Med 91:167-175

Kros JM, Troost D, van Eden CG, van der Werf AJM, Vylings HBM (1988) Oligodendroglioma. A comparison of two grading systems. Cancer 61:2251-2259

Kubo O, Okino T, Kamijo Y, Hamada H, Hirmuro H, Kitamura K (1974) Electron microscopic observations of two cases of monstrocellular sarcoma of the brain in vivo and in vitro (in Japanese). No Shinkei Geka 2:469-475

Kubota I, Hirano A, Yamamoto S (1982) The fine structure of hyaline inclusions in meningioma. J Neuropathol Exp Neurol 41:81-86

Kuhlendahl H, Miltz H, Wullenweber R (1973) Die Astrozytom des Grosshirns. Untersuchirug zur Gruppierung und Prognose. Acta Neurochir 29:151-162

Kumanishi T, Washiyama K, Watabe K, Sekiguchi K (1985) Glial fibrillary acidic protein in medulloblastomas. Acta Neuropathol (Berl) 67: 1-5

Kuo T-T, Hsueh S (1984) Mucicarminophilic histiocytosis. A polyvinylpyrrolidone (PVP) storage disease simulating signet ring carcinoma. Am J Surg Pathol 8:419-428

Kyritsis AP, Tsokos M, Triche T, Chader GJ (1984) Retinoblastoma: Origin from a primitive neuroectodermal cell? Nature 307:471-473

Laerum OD, Bjerkvig R, Steinsvag SK, and de Ridder L(1984) Invasiveness of primary brain tumors. Cancer Metastasis Reviews 3:223-236

Lampson LA, Levy R (1980) Two populations of Ia-like molecules on a human B cell line. J Immunol 125:293-299

Landolt AM, Shibata T, Kleihues P (1987) Growth rate of human pituitary adenomas. J Neurosurg 67:803-806

Landolt AM, Shibata T, Kleihues P, Tuncdogan E (1988) Growth of human pituitary adenomas: Facts and speculations. In: Landolt AM, Heitz PU, Zapf J, Girard J, Del Pozo E (eds) Advances in the Biosciences, Vo1.69. Oxford, New York, Pergamon Press, pp 53-62

Langford LA (1986) The ultrastructure of the ependymoblastoma. Acta Neuropathol (Berl) 71:136-141

Laws ER, Taylor WF, Clifton MB, Okazaki (1984) Neurosurgical management of low grade astrocytomas of the cerebral hemispheres. J Neurosurg 61:665-673

Lee VM-Y, Carden MJ, Trojanowski JQ (1986a) Novel monoclonal antibodies provide evidence for the in situ existence of a nonphosphorylated form of the largest neurofilament subunit. J Neurosci 6:850-858

Lee VM-Y, Carden MJ, Schlaepfer WW (1986b) Structural similarities and differences between neurofilament próteins from five different species as revealed using monoclonal antibodies. J Neurosci 6:2179-2186

Lee VM-Y, Carden MJ, Schlaepfer WW, Trojanowski JQ (1987) Monoclonal antibodies distinguish several differentially phosphorylated states of the two largest rat neurofilament subunits (NF-H and NF-M) and demonstrate their existence in the normal nervous system of adult rats. J Neurosci 7:3474-3488

Leibel SA, Sheline GE (1987) Radiation therapy for neoplasms of the brain. J Neurosurg 66:1-22

Leonard WJ, Depper JM, Uchimaya T, Smith KA, Waldmann TA, Greene WC (1982) A monoclonal antibody that appears to recognize the receptor for human T-cell growth factor: Partial characterization of the receptor. Nature 300: 267-269

Levitt P, Cooper ML, Rakic P (1981) Coexistence of neuronal and glial precursor cells in the cerebral ventricular zone of the fetal monkey: an ultrastructural immunoperoxidase analysis. J Neurosci 1:27-39

Levitt P, Cooper ML, Rakic P (1983) Early divergence and changing proportions of neuronal and glial precursor cells in the primate cerebral ventricular zone. Devel Biol 96:472-484

Libermann TA, Razon N, Bartal AD, Yarden Y, Schlessinger J, Soreq H (1984) Expression of epidermal growth factor receptors in human brain tumors. Cancer Res 44:753-760

Libermann TA, Nusbaum HR, Razon N, Kris R, Lax I, Soreq H, Whittles N, Waterfield MD, Ullrich A, Schlessinger J (1985) Amplification, enhanced expression and possible rearrangement of EGF receptor gene in primary human brain tumors of glial origin. Nature 313:144-147

Lindegaard K-F, Mørk SJ, Eide GE, Halvorsen TB, Hatlevoll R, Solgaard T, Dahli O, and Ganz J (1987) Statistical analysis of clinicopathological features, radiotherapy, and survival in 170 cases of oligodendroglioma. J Neurosurg 67:224-230

Liu HM, Boggs J, Kidd J (1976) Ependymomas of childhood. I. Histological survey and clinicopathological correlation. Child's Brain 2: 92-110

Loeffel SC, Gillespie GY, Mirmiran SA, Miller EW, Golden P, Askin FB, Siegal GP (1985) Cellular immunolocalization of S100 protein within fixed tissue sections by monoclonal antibodies. Arch Pathol Lab Med 109:117-122

Loewenthal A, Flament-Durand J, Karcher D, Noppe M, Brion JP (1982) Glial cells identified by anti-albumin (anti GFA) in human pineal gland. J Neurochem 38:863-865

Louis A (1774) Memoire sur les tumeurs fongueuses de la duremere. Mem Acad Roy Chir Paris 5:1-59

Ludwig CL, Smith MT, Godfrey AD, Armbrustmacher VW (1986) A clinicopathologic study of 323 patients with oligodendrogliomas. Ann Neurol 19:15-21

Ludwin SK, Rubinstein LJ, Russell DS (1975) Papillary meningioma: A malignant variant of meningioma. Cancer 36:1363-1373

Lugaro E (1894) Über die Histogenese der Körner der Kleinhirnrinde. Anat Anz 9:710-713

Lynn JA, Panopio IT, Martin JH, Shaw ML, Race GJ (1968) Ultrastructural evidence for astroglial histiogenesis of the monstrocellular astrocytoma (so-called monstrocellular sarcoma of brain). Cancer 22:356-366

Mabon RF, Svien HJ, Kernohan JW, Craig WMcK (1949) Ependymomas. Proc Staff Meet Mayo Clin 24:65-71

Magdelenat H, Pertuiset BF, Poisson M, Martin PM, Philippon J, Pertuiset B, Buge A (1982) Progestin and oestrogen receptors in meningiomas. Biochemical characterization, clinical and pathological correlations in 42 cases. Acta Neurochirurgica 64:199-2l3

Mai JK, Reifenberger G (1989) Distribution of the carbohydrate epitope 3-fucosyl-N-acetyl-lactosamine (FAL) in the adult human brain. J Chem Neuroanat 1:255-285

Manivel JC, Bulatao I, Wick MR (1988) Meningioma (M): histopathologic and immunohistochemical study. Lab Invest 58:60A

Mannoji H, Takeshita I, Fukui M, Ohta M, Kitamura K (1981) Glial fibrillary acidic protein in medulloblastoma. Acta Neuropathol (Berl) 55:63-69

Mannoji H, Becker LE (1988) Ependymal and choroid plexus tumors: Cytokeratin and GFAP expression. Cancer 61:1377-1385

Marangos PJ, Zomzely-Neurath C, Luk DCM, York C (1975) Isolation and characterization of the nervous system - specific protein 14-3-2 from rat brain. J Biol Chem 250:1884-1891

Mark J (1971) Chromosomal characteristics of neurogenic tumours in adults. Hereditas 68:61-100

Markesbery WR, Challa VR (1979) Electron microscopic findings in primitive neuroectodermal tumors of the cerebrum. Cancer 44:141-147

Markesbery WR, Haugh RM, Young AB (1981) Ultrastructure of pineal parenchyma neoplasms. Acta Neuropathol (Berl) 55:143-149

Markwalder TM, Markwalder RV, Zava DT (1984) Estogren and progestin receptors in meningiomas: Clinicopathological correlations. Clin Neuropharmacol 7:368-374

Martin H, Gottschalk J, Hufnagl P, Szymas J (1986) Bedeutung der Immunhistochemie für die Neuroonkologie. III. Mitteilung: Nachweis des sauren Gliafaserproteins (GFAP) in anaplastischen Astrozytomen und Glioblastomen. Zentralbl allg Pathol Anat 132:209-214

Matsutani M, Nagai M, Takakura K, Sano K (1975) Ultrastructural study of pinealoma in view of its histogenesis. In: Proceedings of the VII International Congress of Neuropathology, Budapest, September 1974. Excerpta Medica, Amsterdam/Akademiai Kiado, Budapest: 413-416

Matsutani M (1976) Ultrastructural study of pinealoma in view of its histogenesis (In Japanese). No To Shinkei 28:41-56

McAllister RM, Isaacs H, Rongey R, Peer M, Au W, Soukup SW, Gardner MB (1977) Establishment of a human medulloblastoma cell line. Int J Cancer 20:206-212

McComb RD, Bourdon MA, Bigner DD (1984) Extracellular matrix (ECM) antigen defined by a monoclonal antibody (MA) raised against the medulloblastoma-derived cell line TE-671: Comparison with the glioma-mesenchymal extracellular matrix (GMEM) antigen and fibronectin (FN) (Abstract). J Neuropathol Exp Neurol 43:336

McGirr SJ, Kelly PJ, Scheithauer BW (1987) Stereotactic resection of juvenile pilocytic astrocytomas of the thalamus and basal ganglia. Neurosurgery 20:447-452

Meis JM, Ordonez NG, Bruner JM (1986) Meningiomas. An immunohistochemical study of 50 cases. Arch Pathol Lab Med 110: 934-937

Michaud J, Gagne F (1983) Microcystic meningioma. Clinicopathologic report of eight cases. Arch Pathol Lab Med 107:75-80

Miller RH, Raff MC (1984) Fibrous and protoplasmic astrocytes are biochemically and developmentally distinct. J Neurosci 4:585-592

Mirimanoff RO, Dosoretz DE, Linggood RM, Ojemann RG, Martuza R (1985) Meningioma: analysis of recurrence and progression following neurosurgical resection. J Neurosurg 62: 18-24

Mirra SS, Miles ML (1980) Unusual pericytic proliferation in a meningotheliomatous meningioma: An ultrastructural study (Abstract) J Neuropathol Exp Neurol 39:376

Mirra SS, Tindall SC, Check IJ, Brynes RK, Moore WW (1983) Inflammatory meningeal masses of unexplained origin. An ultrastructural and immunological study. J Neuropathol Exp Neurol 42:453-468

Moguilewsky M, Pertuiset BF, Verzat BF, Philibert D, Philippon J, Poisson M (1984) Cytosolic and nuclear sex steroid receptors in meningioma. Clin Neuropharmacol 7:375-381

Møller M (1974) The ultrastructure of the human fetal pineal gland. I. Cell types and blood vessels. Cell Tiss Res 152:13-30

Møller M, Ingild A, Bock E (1978) Immunohistochemical demonstration of S-100 protein and GFA protein in interstitial cells of rat pineal gland. Brain Res 140:1-13

Morimura T, Kitz K, Budka H (1989) In situ analysis of cell kinetics in human brain tumors. A comparative immunocytochemical study of S-phase cells by a new in vitro BrdU labeling technique, and of proliferating pool cells by monoclonal antibody Ki-67. Acta Neuropathol (Berl) 77:276-282

Mørk SJ, Rubinstein LJ (1985) Ependymoblastoma. A reappraisal of a rare embryonal tumor. Cancer 55:1536-1542

Mørk SJ, Løken AC. (1977) Ependymoma. A follow-up study of 101 cases. Cancer 40:907-915

Mørk SJ, Halvorsen TB, Lindegaard K-F, Eide GE (1986) Oligodendroglioma. Histologic evaluation and prognosis. J Neuropathol Exp Neurol 45:65-78
Moss R (1987) Free radical - Albert Szent-Gyorgi and the battle over vitamin C. New York, Paragon
Müller E (1903) Zur Ätiologie und Pathologie der Geschwülste des Stirnhirns. Dtsch Z Nervenheilkd 23:378-416
Müller J (1838) Ueber den feineren Bau und die Formen der Krankhaften Geschwülste. Berlin, Reimer
Müller W, Afra D, Schröder R (1977a) Supratentorial recurrences of gliomas. Morphological studies in relation to time intervals with astrocytomas. Acta Neurochir 37:75-91
Müller W, Afra D, Schröder R (1977b) Supratentorial recurrences of gliomas. Morphological studies in relation to time intervals with oligodendrogliomas. Acta Neurochir 39:15-25
Murovic J, Nagashima T, Hoshino T, Edwards MSB, Davis RL (1986) Pediatric central nervous system tumours: A cell kinetics study with bromodeoxyuridine. Neurosurgery 19:900-904
Mushika M, Miwa T, Suzuoki Y, Hayashi K, Masaki S, Kaneda T (1988) Detection of proliferating cells in dysplasia, carcinoma in situ, and invasive carcinoma of the uterine cervix by monoclonal antibody against DNA polymerase a. Cancer 61:1182-1186
Nagashima T, DeArmond SJ, Murovic J, Hoshino T (1985) Immunocytochemical demonstration of S-phase cells by anti-bromodeoxyuridine monoclonal antibody in human brain tumor tissues. Acta Neuropathol (Berl) 67: 155-159
Nagashima T, Murovic JA, Hoshino T, Wilson CB, DeArmond SJ (1986) The proliferative potential of human pituitary tumors in situ. J Neurosurg 64:588-593
Nagashima T, Hoshino T, Cho KG, Edwards MSB, Hudgins RJ, Davis RL (1988) The proliferative potential of human ependymomas measured by in situ bromodeoxyuridine labeling. Cancer 61:2433-2438
Nakagawa Y, Perentes E, Rubinstein LJ (1986) Immunohistochemical characterization of oligodendrogliomas: An analysis of multiple markers. Acta Neuropathol (Berl) 72:15-22
Neidhardt M, Bailey CC, Gnekow A, Kleihues P, Michaelis J, Wellek S (1987) Die Medulloblastom-Therapiestudien MBL 80 und MED 84 der Gesellschaft für Pädiatrische Onkologie (GPO) und der Société International d'Oncologie Pédiatrique (SIOP). Klin Pädiat 199:188-192
Nelson JS, Tsukada Y, Schoenfeld D, Fulling K, LaMarche J, Peress N (1983) Necrosis as a prognostic criterion in malignant supratentorial, astrocytic gliomas. Cancer 52:550-554
Nelson DF, Nelson JS, Davis DR, Chang CH, Griffin T, Pajak TF (1985) Survival and prognosis of patients with astrocytoma with atypical or anaplastic features. J Neuro-Oncol 3:99-103
Niedieck B, Löhler J (1987) Expression of 3-fucosyl-N-acetyl-lactosamine on glia cells and its putative role in cell adhesion. Acta Neuropathol (Berl) 75:173-184
Nielsen SL, Wilson WB (1975) Ultrastructure of a "pineocytoma". J Neuropathol Exp Neurol 34:148-158
Nishizuka Y (1986) Studies and perspectives of protein kinase C. Science 233:305-312
O'Rahilly R, Müller F (1986) The meninges in human development. J Neuropathol Exp Neurol 45:588-608
Ojeda VJ, Jacobsen PF, Papadimitriou JM (1978) Primary cerebral neuroblastoma. Case report with light microscopic study. Acta Neuropathol (Berl) 41:119-124
Ojeda VJ, Stokes BAR, Lee MA, Thomas GW, Papadimitriou JM, Cala LA, Stevens SMB, O'Neill P (1986) Primary cerebral neuroblastomas. A clinicopatho-

logical study of one adolescent and five adult patients. Pathology 18:41-49
Okeda R, Song SJ, Nakajima T, Matsutani M (1984) Pineocytoma - observation of an autopsy case by EM and cell markers. Acta Pathol Jap 34:911-918
Olson JJ, Beck DW, Schlechte J, Loh PM (1986) Hormonal manipulation of meningiomas in vitro. J Neurosurg 65:99-107
Olson JJ, Beck DW, MacIndoe JW, Min-Loh P (1988) Androgen receptors in meningiomas. Cancer 61:952-955
Ostedag CB, Volk B, Shibata T, Burger PC, Kleihues P (1987) The monoclonal antibody Ki-67 as a marker for proliferating cells in stereotactic biopsies of brain tumors. Acta Neurochir 89:117-121
Palmer JO, Kasselberg AG, Netsky MG (1981) Differentiation of medulloblastoma: studies including immunohistochemical localization of glial fibrillary acidic protein. J Neurosurg 55:161-169
Parker JC Jr, Mortara RH, McCloskey JJ (1975) Biological behavior of the primitive neuroectodermal tumors: Significant supratentorial childhood gliomas. Surg Neurol 4:383-388
Pasquier B, Lachard A, Pasquier D, Couderc P, Delpech B, Courel MN (1983) Proteine gliofibrillaire acide (GFA) et tumeurs nerveuses centrales. Etude immunohistochimique d'une sèrie de 107 cas. Ann Pathol 3:203-211
Pearl GS, Takei Y (1981) Cerebellar "neuroblastoma." Nosology as it relates to medulloblastoma. Cancer 47:772-779
Pearl GS, Takei Y, Bakay RAE, Davis P (1985) Intraventricular primary cerebral neuroblastomas in adults: report of three cases. Neurosurgery 16:847-849
Pegram CN, Eng LF, Wikstrand CJ, McComb RD, Lee YL, Bigner DD (1985) Monoclonal antibodies reactive with epitopes restricted to glial fibrillary acidic proteins of several species. Neurochem Pathol 3:119-138
Penfield W (1932) Tumors of the sheaths of the nervous system In: Penfield W (ed) Cytology and Cellular Pathology of the Nervous System. New York, Hoeber, 3:953-990
Perentes E, Rubinstein LJ (1987) Recent applications of immuno-peroxidase histochemistry in human neuro-oncology. Arch Pathol Lab Med 111:796-812
Perentes R, Rubinstein LJ, Herman MM, Donoso LA (1986) S-antigen immunoreactivity in human pineal glands and pineal parenchymal tumors. A monoclonal antibody study. Acta Neuropathol (Berl) 71:224-227
Perentes E, Herbort CP, Rubinstein LJ, Herman MM, Uffer S, Donoso LA, Collins VP (1987) Immunohistochemical characterization of human retinoblastomas in situ with multiple markers. Amer J Ophthalmol 103:647-658
Pertuiset BF, Moguilewsky M, Magdelenat H, Martin PM, Philibert D, Poisson M (1984) Sex steroid receptors in human meningioma and glioma. Progress in Cancer Research and Therapy 31:561-568
Pevet P (1979) Secretory processes in the mammalian pinealocytes under natural and experimental conditions. Prog Brain Res 52:149-194
Pevet P (1981) Ultrastructure of the mammalian pinealocyte. In: Reiter RJ (ed) The Pineal Gland, Vol. 1. Anatomy and Biochemistry. Boca Raton, CRC Press Inc, pp 121-148
Piepmeier JM (1987) Observations on the current treatment of low-grade astrocytic tumors of the cerebral hemispheres. J Neurosurg 67:177-181
Pierce GB, Shikes RH, Fink LM (1978) A problem of developmental biology. Englewood Cliffs, NJ, Prentice-Hall, pp. 68-87
Pierre-Kahn A, Hirsch JF, Roux FX, Renier D, Sainte-Rose C (1983) Intracranial ependymomas in childhood: survival and functional results of 47 cases. Child's Brain 10:145-156
Pilkington GJ, Lantos PL (1982) The role of glutamine synthetase in the diagnosis of cerebral tumors. Neuropathol Appl Neurobiol 8:227-236
Poisson M (1984) Sex steroid receptors in human meningiomas. Clin Neuropharmacol 7:320-324

Romijn HJ, Mud MT, Wolters PS (1976) Electron microscopic evidence of glycogen storage in the dark pinealocytes of the rabbit pineal gland. J Neural Trans 38:231-237

Rorke LB (1983) The cerebellar medulloblastoma and its relationship to primitive neuroectodermal tumors. J Neuropathol Exp Neurol 42:1-15

Rorke LB (1985) Classification and grading of childhood brain tumors. Overview and statement of the problem. Cancer 56:1848-1849

Rorke LB, Gilles FH, Davis RL, Becker LE (1985) Revision of the World Health Organization classification of brain tumors for childhood brain tumors. Cancer 56:1869-1886

Rorke LB (1986) Origin and histogenesis of medulloblastoma. In: Zeltzer PM, Pochedly C (eds) Medulloblastomas in children. New concepts in tumor biology, diagnosis and treatment. New York, Praeger, p. 14-21

Ross GW, Rubinstein LJ (1989) Lack of histopathological cor-relation of malignant ependymomas with postoperative survival. J Neurosurg 70:31-36

Roussy G, Cornil L(1928) A propos de la classification des tumeurs des méninges. Rev Neurol 49:122-125

Rubinstein LJ (1970) The definition of the ependymoblastoma. Arch Pathol 90:35-45

Rubinstein LJ (1972) Cytogenesis and differentiation of primitive central neuroepithelial tumors. J Neuropathol Exp Neurol 31:7-26

Rubinstein LJ (1975) Tumors of the Central Nervous System, 2nd Ser, Fasc.6, Atlas of Tumor Pathology. Washington, Armed Forces Institute of Pathology

Rubinstein LJ (1981a) The cerebellar medulloblastoma: Its origin, differentiation, morphological variants, and biological behavior. In: Vincken PJ, Bruyn GW (eds) Tumors of the brain and skull. Part III. New York, Elsevier, 9:167-194

Rubinstein LJ (1981b) Cytogenesis and differentiation of pineal neoplasms. Human Pathol 12:441-448

Rubinstein LJ (1985a) Editorial. J Neuro-oncol 3:195-196

Rubinstein LJ (1985b) Embryonal central neuroepithelial tumors and their differentiating potential. A cytogenetic view of a complex neuro-oncological problem. J Neurosurg 62:795-805

Rubinstein LJ (1986) Immunological signposts - not markers - in neural tumour differentiation. Neuropathol Appl Neurobiol 12:523-537

Rubinstein LJ (1987a) The correlation of neoplastic vulnerability with central neuroepithelial cytogeny and glioma differentiation. J Neuro-oncol 5:11-27

Rubinstein LJ (1987b) Primitive neuroectodermal tumors (letter). Arch Pathol Lab Med 111:310-311

Rubinstein LJ, Herman MM (1986) The contribution of tissue and organ culture to the differentiating capabilities of the cerebellar medulloblastoma. In: Zeltzer PM, Pochedly C (eds) Medulloblastomas in children. New concepts in tumor biology, diagnosis and treatment. New York, Praeger, pp 37-57

Ruiz-Navarro A, Blanco-Rodriguez A, Gasquez-Ortiz A, Jover-Moyano A (1982) Synaptic ribbons in pinealocytes of castrated rats and rats treated with estradiol. Cell Biol Internat Reports 6:629-633

Russell DS (1944) The pinealoma, its relationship to teratomas. J Clin Path 56:145-150

Russell DS, Rubinstein LJ (1963) Pathology of Tumours of the Nervous System. London, Edward Arnold, pp 173-181

Russell DS, Rubinstein LJ (1988) Pathology of Tumours of the Nervous System, 5th edition. London, Edward Arnold

Russo A, Gianni L, Kinsella T (1984) Pharmacological evaluation of intravenous delivery of 5-bromodeoxyuridine to patients with brain tumors. Cancer Res 44:1702-1705

Salazar OM, Rubin P, Bassano D, Marcial VA (1975) Improved survival of patients with intracranial ependymomas by irradiation: dose selection and field extension. Cancer 35:1563-1573

Salazar OM, Castro-Vita H, VanHoutte P, Rubin P, Aygan C (1983) Improved survival in cases of intracranial ependymoma after radiation therapy. Late report and recommendations. J Neurosurg 59:652-659

Sandberg AA, Morgan R, Berger C, Hecht BK, Hecht F (1984) Chromosome analysis in hematologic disorders: the leukemias. Am J Med 76:971-982

Sano K (1983) Pineal region tumors. Problems in pathology and treatment. Clin Neurosurg 30:59-91

Sano K (1985) Treatment of tumors in the pineal and posterior third ventricular region. International Symposium on Surgery In and Around the Brain Stem and the Third Ventricle, Hannover: 18-23, September 1985

Sano K, Hoshino T, Nagai M (1968) Radiosensitization of brain tumor cells with a thymidine analogue (bromouridine). J Neurosurg 28: 530-538

Sasaki K, Matsumura K, Tsuji T, Shinozaki, F, Takahashi M (1988) Relationship between labeling indices of Ki-67 and BrdUrd in human malignant tumors. Cancer 62:989-993

Sato T, Shimoda A, Takahashi T, Daita G, Goto S, Takamura H, Hirama M. (1980) Congenital cerebellar neuroepithelial tumor with multiple divergent differentiations. Acta Neuropathol (Berl) 50: 143-146

Sato Y, Mukai K, Watanabe S, Goto M, Shimosato Y (1986) The AMEX method. A simplified technique of tissue processing and paraffin embedding with improved preservation of antigens for immunostaining. Am J Pathol 125:431-435

Sayre GP (1956) The concept of grading gliomas of the central nervous system. J Intern Coll Surg 26:440-447

Scanlon PW, Taylor WF (1979) Radiotherapy of intracranial astrocytomas: analysis of 417 cases treated from 1960 through 1969. Cancer 5:301-307

Schaper A (1894) Die morphologische und histologische Entwicklung des Kleinhirn der Teleostiere. Anat Anz 9:489-501

Schaper A (1897) Die frühesten Differenzirungsvorgänge im Centralnervensystem: Kritische Studie und Versuch Einer Geschichte der Entwicklung nervöser Substanz. Arch f Entwicklungsmechn d Organ (Leipzig) 5:81-132

Scheithauer BW, Rubinstein LJ (1979) Cerebral medulloepithelioma. Report of a case with multiple divergent neuroepithelial differentiation. Child's Brain 5:62-71

Scherer HJ (1940) Cerebral astrocytomas and their derivatives. Am J Cancer 40:159-197

Schindler E, Gullotta F (1983) Glial fibrillary acidic protein in medulloblastomas and other embryonic CNS tumors of children. Virchows Arch (Pathol Anat) 398:263-275

Schmidt MB (1902) Ueber die Pacchioni'schen Granulationen und ihr Verhaltniss zu den Sarcomen und Psammomen der Dura mater. Virchows Arch [Pathol Anat] 170:429-464

Schmidt ML, Carden MJ, Lee VM-Y, Trojanowski JQ (1987) Phosphate dependent and independent neurofilament epitopes in the axonal swellings of patients with motor neuron disease and controls. Lab Invest 56:282-294

Schmincke A (1909/1910) Beitrag zur Lehre der Ganglioneurome: Ein Ganglioneurom des Gehirns. Beitr Pathol Anat 47:354-371

Schmincke A (1914) Ein Ganglioneurom des Gehirns. Verh Dtsch Ges Pathol 17:537-543

Schnegg JF, Gomex F, LeMarchand-Beraud T, de Tribolet N (1981) Presence of sex steroid hormone receptors in meningioma tissue. Surg Neurol 15:415-418

Schnitt SJ, Vogel H (1986) Meningiomas: Diagnostic value of immunoperoxidase staining for epithelial membrane antigen. Am J Surg Pathol 10:640-649

Schwab JE, Farber NB, Gottlieb DI (1986) Neurons of the olfactory epithelium in adult rats contain vimentin. J Neurosci 6:208-217

Schwechheimer K, Wiedenmann B Franke WW (1987) Synaptophysin: A reliable broad-range marker for medulloblastomas. Virchows Arch (Pathol Anat) 411:53-59

Seeger RC, Danon YL, Rayner SA, Hoover F (1982) Definition of a Thy-1 determinant on human neuroblastoma, glioma, sarcoma, and teratoma cells with a monoclonal antibody. J Immunol 128:983-989

Shapiro WR, Green SB, Burger PC, et al. (1989) Randomized trial of three chemotherapy regimens (BCNU vs BCNU alternating with procarbazine vs BCNU plus hydroxyurea alternating with procarbazine plus VM-26) and two radiotherapy regimens (whole brain vs partial whole brain plus coned-down radiotherapy) in the postoperative treatment of malignant glioms: Brain Tumor Cooperative Group (BTCG) Trial 8001. J Neurosurg (in press)

Shaw EG, Evans RG, Scheithauer BW, Ilstrup DM, Earle JD (1987) Postoperative radiotherapy of intracranial ependymoma in pediatric and adult patients. Int J Radiation Oncol Biol Phys 13:1457-1462

Shaw EG, Daumas-Duport C, Scheithauer BW, Gilbertson DT, O'Fallon JR, Earle JD, Laws ER, Okazaki H (1989) Controversies in the management of low grade supratentorial astrocytomas; defining the role of radiation therapy. J. Neurosurg (in press)

Shibata T, Burger PC, Kleihues P (1988) Ki-67 immunoperoxidase stain as marker for the histological grading of nervous system tumours. Acta Neurochir Suppl 43:103-106

Shin WY, Laufer H, Lee YC, Aftalion B, Hirano A, Zimmerman HM (1978) Fine structure of a cerebellar neuroblastoma. Acta Neuropathol (Berl) 42:11-13

Shuangshoti S (1980) Melanotic mucin-producing neuroepithelial neoplasm of mesencephalon with consideration of similar tumours in different locations. J Neurol Neurosurg Psychiatr 43:810-817

Simpson D (1987) The recurrence of meningiomas after surgical treatment. J Neurol Neurosurg Psychiatr 20:22-39

Skullerud K, Løken AC (1974) The prognosis in meningiomas. Acta Neuropath (Berl) 29:337-344

Smith MT, Ludwig CL, Godfrey AD, Armbrustmacher VM (1983) Grading of oligodendrogliomas. Cancer 52:2107-2114

Smith TW, Davidson RI (1984) Medullomyoblastoma: A histologic, immunohistochemical, and untrastructural study. Cancer 54:323-332

Sobel RA, Trice JE, Nielsen SL et al (1981) Pineoblastoma with ganglionic and glial differentiation: Report of two cases. Acta Neuropathol (Berl) 55:243-246

Sobel RA, Michaud J (1985) Microcystic meningioma of the falx cerebri with numerous palisaded structures: An unusual histological pattern mimicking schwannoma. Acta Neuropathol (Berl) 68:256-258

Stage WS, Stein JJ (1974) Treatment of malignant astrocytomas. Radiology 120:7-18

Stam FC, van Alphen HAM, Boorsma DM (1980) Meningioma with conspicuous plasma cell components. A histopathological and immunohistochemical study. Acta Neuropathol (Berl) 241-243

Stefanko SZ, Manschot WA (1979) Pinealoblastoma with retinoblastomatous differentiation. Brain 102:321-332

Stowens D, Lin T-H (1974) Melanotic progonoma of the brain. Human Pathol 5:105-113

Svien HJ, Mabon RF, Kernohan JW, Adson AW (1949) Astrocytomas. Proc Staff Mtg. Mayo Clinic 24:54-64

Swarz JR, del Cerro M (1977) Lack of evidence for glial cells originating from the external granular layer in mouse cerebellum. J Neurocytol 6:241-250

Szymas J (1986) Das saure Gliafaserprotein. Ein wertvolles Hilfsmittel in der neuropathologischen Diagnostik. Zentralbl allg Pathol Anat 131:3-19

Szymas J, Reifenberger G, Wechsler W (1987) Leu-Ml immunoreactivity in the human brain - Discrimination between different tumors and between neoplastic and brain tissue. Naturwissenschaften 74:188-189

Szymas J, Biczysko W, Gabryel P, Morkowski S (1987a) Medulloblastoma: histological evaluation and prognosis. Child's Nerv Syst 3:74-80

Szymas J, Hossmann K-A, Weber F, Oschlies U (1988a) Glial fibrillary acidic protein in human glioblastoma. (in press)

Szymas J, Hossmann K-A, Weber F, Oschlies U (1988b) Immunoelectron microscopic analysis of lacto-N-fucopentaose III in brain tumours and human brain tissue. (in preparation)

Tang TT, Harb JM, Mørk SJ, Sty JR (1985) Composite cerebral neuroblastoma and astrocytoma. A mixed central neuroepithelial tumor. Cancer 56:1404-1412

Temple S, Raff MC (1985 Differentiation of a bipotential glial progenitor cell in single cell microculture. Nature 313:223-225

Theaker JM, Gatter KC, Esiri MM, Fleming KA (1986) Epithelial membrane antigen and cytokeratin expression by meningiomas: an immunohistological study. J Clin Pathol 39:435-439

Tijssen CC, Halprin MR, Endtz LJ (eds) (1982) Familial Brain Tumors: A Commented Registry. The Hague (Leyden), Martinus Nijhoff

Townsend JJ, Seaman JP (1986) Central neurocytoma - a rare benign intraventricular tumor. Acta Neuropathol (Berl) 71:167-170

Triche TJ (1986) Neuroblastoma - biology confronts nosology. Arch Pathol Lab Med 110:994-996

Triche TJ (1987) Primitive neuroectoermal tumors (letter). Arch Pathol Lab Med 111:311-312

Trojanowski J (1987) Neurofilament proteins and human nervous system tumors. J Histochem Cytochem 35:999-1003

Tsokos M, Kyritsis AP, Chader GJ, Triche TJ (1986) Differentiation of human retinoblastoma in vitro into cell types with characteristics observed in embryonal or mature retina. Am J Pathol 123:542-552

VandenBerg SR, Herman MM, Ludwin SK, Bignami A (1975) An experimental mouse testicular teratoma as a model for neuroepithelial neoplasia and differentiation: I. Light microscopic and tissue and organ culture observations. Am. J Pathol 79:147-168

VandenBerg SR, Mary EE, Rubinstein LJ, Herman MM, Perentes E, Vinores SA, Collins VP, Park TS (1987a) Desmoplastic supratentorial neuroepithelial tumors of infancy with divergent differentiation potential (desmoplastic infantile ganglioglioma). A report on 11 cases of a distinctive embryonal tumor with favorable prognosis. J Neurosurg 66:58-71

VandenBerg SR, Herman MM, Rubinstein LJ (1987b) Embryonal central neuroepithelial tumors: current concepts and future challenges. Cancer Metast Rev 5:343-364

Varakis JN, Zu Rhein GM (1976) Experimental pineocytoma of the Syrian hamster induced by a human papovavirus (JC). A light and electron microscopic study. Acta Neuropathol (Berl) 35:243-264

Velasco ME, Ghobrial MW, Ross ER (1985) Neuron-specific enolase and neurofilament protein as markers of differentiation in medulloblastoma. Surg Neurol 23: 177-182

Virchow, R (1864) Die krankhaften Geschwulste. Berlin, Hirschwald 2:343

Vollrath L (1973) Synaptic ribbons of a mammalian pineal gland. Circadian changes. Z Zellforsch 145:171-183

Vollrath L (1981) The pineal organ. Berlin, Heidelberg, Springer-Verlag

Wada C, Kurata A, Hirose R, Tazaki Y, Kan S, Ishihara Y, Kameya T (1986) Primary leptomeningeal ependymoblastoma. J Neurosurg 64:968-973

Yagishita S, Itoh Y, Chiba Y, Kuwana N (1982) Morphological investigations on cerebellar "neuroblastoma" group. Acta Neuropathol (Berl) 56:22-28

Yamada K, Kondo T, Yoshioka M, Oami H (1980) Cytogenetic studies in twenty human brain tumors: association of no. 22 chromosome abnormalities with tumors of the brain. Cancer Genet Cytogenet 2:293-307

Yoshii Y, Maki Y, Tsuboi K, Tomono Y, Nakagawa K, Hoshino T (1986) Estimation of growth fraction with bromodeoxy-uridine in human central nervous system tumors. J Neurosurg 65:659-663

Young S, Rothbard J, Parker PJ (1988) A monoclonal antibody recognizing the site of limited proteolysis of protein kinase C. Eur J Biochem 173:247-252

Zankl H, Zang KD (1972) Cytological and cytogenetical studies on brain tumors. IV. Identification of the missing G chromosomes in human meningiomas as no. 22 by fluorescence technique. Humangenetik 14:167-169

Zarbo RJ, Meis JM, Ho K-L, Hatfield H, Crissman JD (1988) Immunocharacterization of meningiomas: comparison of frozen and formalin fixed tumors. Lab Invest 58:108A

Zuber P, Hamou M-F, de Tribolet N (1988) Identification of proliferating cells in human gliomas using the monoclonal antibody Ki-67. Neurosurgery 22:364-368

Zülch KJ (1939) Über die geschichtliche Entwicklung und den heutigen Stand der Klassifikation der Hirngeschwülste im Jugendalter. Zentralb Neurochir 4:251- 272; 325-335

Zülch KJ (1940) Hirngeschwülste im Jugendalter. Zentrabl Neurochir 5: 238-274

Zülch KJ (1953) Über die primären Hirnsarkome. (Mit beson-derer Berücksichtigung der monstrocellulären Formen) Arch Int Studi Neurol 2:1-35

Zülch KJ (1956) Biologie und Pathologie der Hirngeschwülste. In: Olivecrona H, Tönnis W (eds) Handbuch der Neurochirurgie, Vol.3. Berlin, Springer

Zülch KJ (1965) Brain tumors - their biology and pathology, 2nd edition. New York, Springer

Zülch KJ, Wechsler W (1968) Pathology and Classification of Gliomas. Progr Neurol Surg, Vol.2. Basel, New York, Karger, pp 1-84

Zülch KJ (1975) Atlas of Gross Neurosurgical Pathology. Berlin, Heidelberg, New York, Springer

Zülch KJ (1979) Histologic Classification of Tumours of the Central Nervous System. International Histological Classification of Tumours, No. 21, World Health Organization, Geneva

Zülch KJ (1980) Principles of the new World Health Organization (WHO) classification of brain tumors. Neuroradiology 19:59-66

Zülch KJ (1986) Brain Tumors. Their biology and pathology, 3d ed. Springer, Berlin, Heidelberg, New York, Tokyo

Subject Index

MIX
Papier aus verantwortungsvollen Quellen
Paper from responsible sources
FSC® C105338

If you have any concerns about our products,
you can contact us on
ProductSafety@springernature.com

In case Publisher is established outside the EU,
the EU authorized representative is:
Springer Nature Customer Service Center GmbH
Europaplatz 3, 69115 Heidelberg, Germany

Printed by Libri Plureos GmbH
in Hamburg, Germany